Teacher's Edition

Laboratory Manual

Prentice Hall
Earth Science

PEARSON

Prentice Hall

Boston, Massachusetts
Upper Saddle River, New Jersey

CONTENTS

Introduction to the Teacher   T3

Guidelines for Laboratory Safety   T6

Laboratory Investigations Materials and Equipment   T10

Suppliers of Laboratory Materials and Equipment   T16

Laboratory Manual
Teacher's Edition

Prentice Hall *Earth Science*

Acknowledgments
Page **DB2 (Crystals A, B, C and E),** Jeff Scovil/Scovil Photography; **DB2 (Crystal D),** Charles D. Winters/Photo Researchers, Inc.; **DB3 (Crystal F),** Carolina Biological/Visuals Unlimited; **DB3 (Crystal G),** Breck Kent/Earth Scenes; **DB3 (Crystal H),** José Manuel Sanchis Calvete/CORBIS; **DB3 (Crystal I),** Gary Retherford/Photo Researchers, Inc.; **DB3 (Crystal J),** Jeff Scovil/Scovil Photography; **DB6–DB7,** David Sandwell, Scripps Institution of Oceanography; **DB13,** Carrie Gowran; **DB16–DB17,** Unisys Corporation.

Copyright © by Pearson Education, Inc., publishing as Pearson Prentice Hall, Boston, Massachusetts 02116. All rights reserved. Printed in the United States of America. This publication is protected by copyright, and permission should be obtained from the publisher prior to any prohibited reproduction, storage in a retrieval system, or transmission in any form or by any means, electronic, mechanical, photocopying, recording, or likewise. For information regarding permission(s), write to: Rights and Permissions Department, One Lake Street, Upper Saddle River, New Jersey 07458.

ISBN 0-13-125900-8

5 6 7 8 9 10   10 09 08 07 06

Introduction to the Teacher

Earth science is an exciting area of study for the high school student, and the Prentice Hall *Earth Science Lab Manual* is designed to capture this excitement through a variety of interesting and informative activities. In addition to reinforcing key concepts and scientific terms, these activities help students understand and appreciate the processes involved in scientific experimentation.

The Lab Manual contains 60 Laboratory Investigations directly related to the information presented in Prentice Hall *Earth Science*. It also contains 4 Laboratory Skills activities that give students an opportunity to practice and master specific laboratory skills, which include applying the scientific method and identifying errors. In addition, the Lab Manual includes worksheets for the end-of-chapter labs contained in Prentice Hall *Earth Science*.

This Teacher's Edition provides the information you will need to perform the investigations and activities with your students. It includes answers to all investigations and activities, guidelines for laboratory safety, comprehensive lists of laboratory equipment and materials, and a number of annotations that will help you organize and stock your biology laboratory and facilitate your teaching.

Laboratory Investigation Format

The investigations and activities in the Lab Manual are designed to strengthen students' laboratory, critical thinking, and science process skills and to provide a practical application of material presented in the student textbook. The easy-to-follow format of each Laboratory Investigation allows students to complete the investigations on their own, perhaps after an overview and brief explanation from you. This gives you an opportunity to provide the necessary help to those individuals or groups of students who require teacher assistance.

Each Laboratory Investigation has the following format:

Title A brief title presents the topic explored in the investigation. To the right of the title are teacher annotations regarding the time required for the investigation and the textbook pages in Prentice Hall *Earth Science*.

Introduction The Introduction section develops relevant concepts or relates the Laboratory Investigation to specific concepts discussed in the student textbook. In a few cases, this section may also provide new concepts and terms specific to the investigation.

Problem Each Laboratory Investigation challenges the student by introducing a problem in the form of a question or series of questions. Students should be able to solve the problem(s) upon successful completion of the Laboratory Investigation.

Pre-Lab Discussion Asking them to identify the role of certain materials or the reasons for specific steps in the Procedure, questions in this section prepare students for the active understanding needed to take full advantage of the Laboratory Investigation. Questions in this section may also highlight safety procedures to which students should pay careful attention.

Materials This section lists all materials required to conduct the investigation. The quantity of material for each investigation is indicated for individual students, pairs of students, or groups of students. Teacher annotations on how to prepare materials specific to the investigation are found in or near the Materials section. The necessary equipment and materials for all of the Laboratory Investigations are listed alphabetically and by category on pages T10 to T16.

Safety If a Laboratory Investigation requires specific safety precautions, students are alerted to that fact in the special Safety section as well as in the Procedure. The Safety section is intended to make students aware of potential hazards before the investigation is begun, thereby decreasing the risk of accidents.

Procedure This section provides a detailed step-by-step outline of the Laboratory Investigation procedure. Diagrams are included where necessary to further explain a technique or illustrate an experimental setup.

The Procedure contains a number of symbols and instructions that guide students as the lab is performed. Safety symbols, which appear next to certain steps in the Procedure, signal students to begin following specific safety precautions for an indicated step in the procedure and all following steps. **CAUTION** statements within the steps of the Procedure warn students of possible hazards and indicate how accidents can be avoided. **Notes** in the Procedure direct students' attention to special directions or techniques. At the end of some steps in the Procedure, students may be instructed to record data or answer questions.

In keeping with the traditional scientific method, observations are recorded in each investigation. Observations are often recorded by filling in data tables, graphing data, labeling diagrams, and drawing observed structures, as well as answering general questions.

In the Teacher's Edition, annotations to the questions in Procedure and Analysis and Conclusions are printed in red on the pages corresponding to the student pages. These annotations include answers to objective questions, sample drawings and diagrams, and anticipated student data when these can be predicted.

Analysis and Conclusions Questions in this section are designed to assist students in answering the investigation's Problem, relating the investigation to concepts learned in the textbook, and drawing conclusions about the results of their investigation. Using data gathered during the investigation and knowledge gained from the textbook and Introduction, students are asked to analyze and interpret their experimental results. Many questions emphasize possible applications of the experiment and allow students to relate the investigation to real-life situations.

Go Further Most Laboratory Investigations conclude with a section entitled Go Further. This section suggests additional activities that may be used to enrich or supplement the investigation. The Go Further activities may also be used as alternatives to the Laboratory Investigation. Complete instructions are included so that individual students can perform the Go Further activities without additional teacher help.

Design Your Own Experiment Labs The Laboratory Manual also contains investigations that ask students to develop their own experiments. In these activities, the Pre-Lab Discussion helps guide students in planning their experiments. A list of **Suggested Materials** follows. In place of a set Procedure, a section titled **Design Your Experiment** leads students through the planning and performance of their own experiments.

Modifying the Laboratory Investigations

School schedules often do not permit extended laboratory periods. The Laboratory Manual provides a number of options to help you circumvent the difficulties associated with having limited laboratory time. First, most investigations require only 30 to 40 minutes and thus can be completed within a single class period. Second, the investigations are often divided into two or more parts. This provides flexibility by allowing you to select the parts of the investigation that best suit your needs, objectives, and laboratory situation. In addition, you can often schedule the parts of an investigation to be performed during different class periods, as your schedule permits. Finally, a number of the investigations have annotations that suggest specific ways in which the investigation may be modified for a shorter laboratory period.

Correlation Between Textbook Chapters and Laboratory Investigations

The Laboratory Manual is designed to accompany Prentice Hall *Earth Science* in the presentation of a comprehensive Earth science program for high school students. When used in conjunction with the textbook, the Lab Manual reinforces, expands, and enhances the student's experiences of reading the textbook and participating in classroom discussions. Although the investigations are numbered to correspond to chapters in Prentice Hall *Earth Science*, they can be used with any high school Earth science program.

Guidelines for Laboratory Safety

Safety should be an integral part of the planning, preparation, and implementation of a laboratory program. Both the science teacher and the student are responsible for creating and maintaining an enjoyable, instructional, and safe environment in the science laboratory.

A number of general safety concerns are discussed. You should also refer to pages x through xiii in the Student Edition of the Laboratory Manual for a detailed discussion of safety rules and procedures. You may want to expand or modify the safety guidelines, procedures, and rules suggested here according to local and state government regulations and policies, as well as school regulations and policies. The school administration should be able to provide you with specifics on local safety requirements.

General Safety Considerations

Emphasis on proper safety precautions for each laboratory investigation is an essential part of any pre-laboratory discussion. Prior to each investigation, demonstrate the proper use of the required equipment. Demonstrate any potentially hazardous procedure used in that investigation. Always wear the required safety protective devices during the demonstrations and the investigations. If students are required to wear safety goggles, you and any visitors to the class must also wear them.

During an investigation, move about the laboratory to keep constant watch for potentially dangerous situations. Behavior that is inappropriate to a laboratory situation should be curtailed immediately. Wild play and practical jokes are forbidden in the laboratory. Once students realize that the practice of safety is a required part of the course, they will accept a serious approach to laboratory work.

Any laboratory investigation a student performs should have your prior approval. Students should never work in the laboratory without adult supervision. At the conclusion of the lab investigation, cleanup should follow authorized guidelines for waste disposal. The laboratory should be restored to a safe condition for the next class.

Classroom Organization

Furniture and equipment in the laboratory should be arranged to minimize accidents. Assign students to laboratory stations. Each station should be equipped with a flat-topped table and laboratory bench. Do not use desks with slanted tops. Provide several locations where students can obtain needed supplies. Control traffic flow in the room to prevent collisions between students who are carrying or handling equipment. Tell students to leave their personal property in a designated location, away from the laboratory stations. Do not use the floor and benches for storage area. Stress that good housekeeping is important in maintaining safe laboratory conditions. Students should keep all laboratory work areas clean. Unnecessary papers, books, and equipment should be removed from working areas.

Be sure that water faucets, hot plates, gas outlets, and alcohol or Bunsen burners are turned off when not in use.

Safety Equipment

Any classroom where laboratory investigations are done should contain at least one each of the following pieces of safety equipment: (1) fire extinguisher, (2) fire blanket, (3) fire alarm, (4) phone or intercom to the office, (5) eyewash station, (6) safety shower, (7) safety hood, and (8) first-aid kit. If any of these basic pieces of safety equipment are not available, you may need to modify your laboratory program until the situation is remedied.

Make sure students know the location and proper use of all safety equipment. Where appropriate and practical, have students handle or operate the equipment so that they become familiar with it. Make sure all safety equipment is in good working order. All malfunctions should be promptly reported in writing to the proper school or district administrator.

Fire Equipment At the beginning of the school year, you may wish to give each student the opportunity to actually operate a fire extinguisher, as the sound and action of a CO_2 fire extinguisher can be quite alarming to those who never used one. You may also want to have students practice smothering imaginary flames on one another with the fire blanket.

Eyewash Station The eyewash station should be used if chemicals are splashed onto the face or eyes. The exposed area should be left in the running water for 15 minutes.

Safety Shower The shower is used when chemicals have been spilled on a student's body or clothing. The student should remove contaminated clothing and stand under the shower until the chemical is completely diluted. Have a bathrobe or some other type of replacement clothing handy in case the student's clothing must be removed.

You may want to set up one or two spill kits in your laboratory. The contents of a spill kit are used to neutralize chemicals such as acids and bases so that they can be cleaned up more easily. Baking soda (sodium bicarbonate) can be used to neutralize acids. Vinegar (acetic acid) can be used to neutralize bases. Commercial spill kits for acids, bases, and a number of other chemicals are available from supply houses.

Safety Hood Use a safety hood whenever students are working with volatile or noxious chemicals. Make sure that the room is well-ventilated when students are using any kind of chemicals or are working with preserved specimens. Warn students of the flammability and toxicity of various chemicals.

First-Aid Kit A typical first-aid kit contains an assortment of antiseptics, bandages, gauze pads, and scissors. Most also contain simple instructions for use. Be sure to read the instructions if you are not familiar with basic first-aid procedures. A first-aid kit should be taken on all field trips. For field trips, you may wish to add such items as a bee-sting kit, meat tenderizer, tweezers, and calamine lotion. Do not dispense medication (including aspirin).

Cleanup

Before beginning an investigation, instruct students in the proper cleanup procedures. Mark certain containers for the disposal of wastes and the collection of soiled glassware and equipment. Have students dispose of broken glassware in a separate trash container. Before the end of the laboratory period, have students unplug microscopes and other pieces of equipment and put them away in their proper location. Have students wash glassware, wipe up spills, and do whatever else is necessary to clean up their work area. At the conclusion of the laboratory investigation, the room should be restored to a clean and safe condition for the next class. You may wish to institute a policy of not dismissing the class until the laboratory area meets with your approval.

Preparations and the Storage Room

Reagents stored in the stockroom should be clearly labeled and stored safely. Take inventory of reagents frequently and keep up-to-date records of their use. Check local and state regulations for maximum permissible amounts of reagents allowed in school. In case of fire or vandalism, inform the authorities of possible hazards to the community. Keep all chemicals in a locked storage area that is accessible only to you or individuals under your direct supervision.

Some chemicals are incompatible and should be stored separately. Check local and state laws for regulations on storage of flammable liquids. The National Fire Protection Association recommends that flammable liquids be stored in vented, flame-resistant cabinets. Store large containers near floor level. Make sure that storage shelves have a raised lip at the front to prevent containers from sliding forward.

Hazardous Materials

Some reagents can be explosive and should not be on the premises. If found, they should be removed by trained fire or police bomb squads or by other qualified officials.

If you have doubts about the hazards of any reagent in the stockroom, contact an appropriate agency (NIOSH or a local health agency).

Known carcinogens commonly found in school science laboratories include the following: arsenic powder, arsenic trichloride, arsenic pentoxide, arsenic trioxide, asbestos, benzene, benzidine, chromium powder, formaldehyde, lead arsenate, sodium arsenate.

Probable carcinogens include the following: acrylonitrile, cadmium chloride, cadmium powder, cadmium sulfate, carbon tetrachloride, chloroform, ethylene oxide, nickel powder.

Exercise great care in using refrigerators. Never store flammable liquids in a refrigerator unless it is explosion-proof. Do not store food where microbial cultures are stored. Clean refrigerators frequently and safely discard old material.

Laboratory Glassware

Probably the most common school laboratory accidents involve cuts from chipped or broken glassware and burns from hot glassware. Discard any glassware that has a crack or chip. Use only borosilicate glassware. Fire-polish the ends of glass tubing. Allow hot glassware to cool on a hot pad for several minutes before picking it up. If an accident should happen, first aid for minor cuts and burns is immersion in cool running water. For cuts that are bleeding heavily, apply pressure with folded toweling or gauze. Call a health professional immediately.

To insert glass tubing into a stopper, lubricate the stopper hole and the tubing. Wrap the tubing in several layers of toweling and gently work the tubing into the stopper, using a twisting motion and keeping the hands as close together as possible. Wear heavy gloves. Remove the tubing in the same manner as soon as possible. Tubing that is stuck is nearly impossible to remove without cutting the stopper.

Measuring small amounts of liquids with pipettes is common in investigations. But never pipette by mouth. Use rubber suction bulbs designed for use with pipettes or pipette fillers.

Field Studies

Before taking the students on a field study, examine the area for possible safety hazards. Look for terrain or water hazards and poisonous plants and animals. Obtain the necessary written permission from parents and school authorities. Instruct students on proper dress and behavior. Make sure that students are thoroughly familiar with the investigations they are to conduct. If students are to form small groups, decide in advance when and where they will reassemble. Do not allow any student to travel alone.

Identify any students who have special health problems, especially allergies. Alert these students to potential hazards. Be sure they are adequately prepared to deal with emergencies.

Guidelines for Safe Disposal of Laboratory Wastes

Every effort should be made to recover, recycle, and reuse materials used in the laboratory. When disposal is required, however, specific procedures should be followed in order to ensure that your school complies with local, state, and federal regulations.

1. Discard only dry paper into ordinary wastebaskets.
2. Discard broken glass into a separate container clearly marked "For Broken Glass Only."
3. Acidic or basic solutions need to be neutralized before disposal. Slowly add dilute sodium hydroxide to acids and dilute hydrochloric acid to bases until pH paper shows that they are no longer strongly acidic or basic. Then flush the solutions down the drain with a lot of water.
4. Before each investigation, instruct your students concerning where and how they are to dispose of chemicals that are used or produced during the investigation. Specific teacher notes addressing disposal are provided on each lab as appropriate.
5. Keep each excess or used chemical in a separate container; do not mix them. This allows for possible recycling or reuse. It also eliminates unexpected reactions or the need for expensive separation by a contractor if the wastes must be disposed of professionally.
6. Only nonflammable, neutral, nontoxic, nonreactive, and water-soluble chemicals should be flushed down the drain.

Laboratory Investigations Materials and Equipment

Note: Safety equipment has not been listed. A laboratory apron, safety goggles, and plastic or heat-resistant gloves should be worn when required.

Item	Quantity Per Group	Laboratory Investigation
Aluminum		
foil	15 cm × 15 cm	4B
pan, large	1	5
Atlas	1	11
Balance, laboratory	1	1A
Balloon, large, round	1	8A
Beaker		
150-mL	2	Intro B
250-mL	4	4A
500-mL	1	Intro A, 4B
Beam, wood, 30–40 cm long	1	8B
Blocks, wooden, various heights	3	5
Board, wooden, 3-cm thick	1	4B
Bottle		
1-L plastic	1	5
2-L plastic, transparent, with caps	2	21
2-L plastic	2	4B
plastic with spray pump	1	4A
Bowl, medium-sized, clear, plastic	1	6B
Box, flat with lid	1	12
Brad, metal	1	2
Brick, small, thin	1	8B
Bucket	1	5
Burner, laboratory	1	4B, 21
Calculator	1	1A, 7, 22, 23
Candy, hard	5 pieces	6B
Cardboard		
heavy corrugated	50 cm × 60 cm	6B, 14, 23
9 in. × 11 in.	1 piece	14
stiff, white, 20 cm × 10 cm	1 piece	2
stiff square, 35 cm wide	1	25
thin, 8 1/2 in. × 11 in.	1 piece	14
Compass, drawing	1	24
Cord, heavy	1 spool	8B
Cough drop	5	6B
Dowel, wooden	1	8B
Dropper	1	3
Epsom salts	800 mL	6B
Flask, Erlenmeyer, 250-mL	2	4B
Funnel	1	8A
Gauze, wire, 15 cm × 15 cm	1	4B
Glass tubing, bent at right angles	2 pieces	4B

Earth Science Lab Manual

Gloves		
heat-resistant	1 pair	4B, 21
plastic	1 pair	14
Glue	1 bottle	14
all-purpose	1 bottle	6B, 8A
clear, quick-drying	1 bottle	2
Graduated cylinder		
50-mL	1	Intro A, 18
100-mL	2	Intro B, 4A, 4B
250-mL	1	8A
large	1	1A, 6B
Hammer	1	8B
Hand drill	1	8B
Hand lens	1	3
Hand saw	1	8B
Hole punch	1	2
Hydrochloric acid, 1M, dilute	1 bottle	3
Ice		
crushed	500 mL	4B
cube	1	Intro B
Index card, 10 cm × 15 cm	2	24
Jar		
glass, 1-L	1	6B
glass, gallon	1	18
Knife, serrated	1	6B
Knitting needle, large, metal	1	21, 25
Lamp, gooseneck with 100-W bulb	1	21
Map		
North America	1	11
topographic, enlarged photocopy	1	1B
Masses, identical small	2	25
Match	1	4B
safety	2	18, 21
Measuring cup	1	8A
Meter stick	2	Intro A, 1A, 22, 24
Mineral oil	200 mL	4A
Modeling clay	300 g	1B, 6B, 21
Nail	several	8B
Nickel	1	1A
Paint, tempera, several colors	4–5 jars	14
Paper		
filter	1 sheet	8A
graph	1 sheet	17B
tracing	1 sheet	14
white, plain	3 sheets	6A, 23
Paper clip	1	1A
Paper cup	1	1A
Paper towel	1 roll	3, 4A, 6B

Item	Quantity	Lab
Pebble, pea-sized	several	4A
Pen,		
glass-marking	1	4B
nonpermanent, fine-lined marking	1	1B
red	1	3
Pencil	1	6A
colored	several	11, 20A, 21, 23
blue	1	19
glass-marking	1	Intro B
red	1	3, 19
sharpened, with soft lead	1	8B
with eraser	1	25
Penny	100	Intro A, 12
Plank, wood, 2.5-cm thick	1	8B
Plastic cup, 500-mL capacity	2	6B
Plastic strip, thin, clear, rigid, 15 cm × 2 cm	1 strip	2
Plastic tubing, 40-cm long	2 pieces	4A, 4B
Protractor	1	5, 9, 11, 22
small, thin	1	2
Pushpin	2	23, 25
Reinforcement, gummed	2	2
Ring stand, with ring	1	4B
Rock		
igneous	1	3
metamorphic	1	3
sedimentary	1	3
small	1	1A
Rubber band, 15-cm circ.; 0.5-cm width	1	18
Ruler	1	9, 11, 17B
metric	1	1A, 1B, 2, 5, 6A, 7, 8B, 23, 24, 25
Salt, table	800 mL	6B
Saltwater solution	100 mL	4B
Sand		
clean, dry, medium-grained	500 g	8A
play	1 bag	5
Scissors	1	2, 5, 8A, 14, 24
Screw	several	8B
Shoebox, transparent with lid	1	1B
Skewer, wooden	1	6B
Soap, dishwashing	1 bottle	4A
Spoon, large	1	6B
Sprinkling can	1	5
Stirring rod	1	6B
Stopper, one-hole rubber	2	8A
with glass tubing inserted	1	4B
Storage bag, plastic freezer, 26 cm × 26 cm	1	18
Straight edge	1	17B
Straw, drinking		
large	1	8A
plastic	4	5
String	30-cm piece	11, 23
cotton	60-cm piece	6B
Sugar	800 mL	6B
Sugar crystal	5	6B

Tap water		
cold	80 mL	4A, 18
hot	40 mL	4A, 6B, 18
room-temperature	1000 mL	Intro A, 8A
Tape		
adding machine	1 roll	8B
cellophane	1 roll	1B, 2, 23
duct	1 roll	4B, 5, 8B
masking	1 roll	1B, 24, 25
Tape measure, metric	1	1A
Thermometer, Celsius, non-mercury	2	Intro A, Intro B, 21
Tongs	1	6B
Trash bag, heavy-duty	1	5
Vacuum pump	1	8A
Vegetable oil	200 mL	4A
Vice grips	1	8B
Vinegar, white	250 mL	6B
Washer, metal, small	2	6B
Watch or clock	1	Intro B, 21
with second hand	1	4A, 25
Wire, heavy	1 spool	8B
Wire cutters	1	8B

Materials Inventory

Equipment

Aluminum foil
Atlas

Balance, laboratory
Balloon
Beam, wood
Block, wooden
Board, wooden
Bottle
    1-L plastic
    2-L plastic
    plastic with spray pump
Bowl, clear plastic
Box, flat with lid
Brad, metal
Brick
Bucket
Burner, laboratory

Calculator
Cardboard
    heavy corrugated
    stiff, white
    thin
Compass, drawing
Cord, heavy

Dowels, wooden

Funnel

Gauze, wire
Gloves
    heat-resistant
    plastic
Glue
    all-purpose
    clear, quick-drying

Hammer
Hand drill
Hand lens
Hand saw
Hole punch

Index card

Knife, serrated
Knitting needle, metal

Lamp, gooseneck with
    100-W bulb

Map
    North America
    topographic, enlarged
        photocopy
Masses, identical
Matches
    safety
Measuring cup
Meter stick
Modeling clay

Nail
Nickel

Paint, tempera
Pan
Paper
    filter
    graph
    tracing
    white, plain
Paper clip
Paper cup
Paper towel
Pebble
Pen
    glass-marking
    nonpermanent, fine-lined,
        marking
    red
Pencil
    colored
    blue
    glass-marking
    red
    sharpened, with soft lead
    with eraser
Penny
Plank, wood
Plastic cup, 500-mL capacity
Plastic strip

Plastic tubing
Protractor
    small, thin
Pushpin

Reinforcement, gummed
Ring stand, with ring
Rock
    igneous
    metamorphic
    sedimentary
    small
Rubber band
Ruler
    metric

Sand
    medium-grained
    play
Scissors
Screw
Shoebox, transparent with lid
Skewer, wooden
Soap, dishwashing
Spoon
Sprinkling can
Stirring rod
Stopper, one-hole rubber
    with glass tubing inserted
Storage bag, plastic freezer
Straight edge
Straw, drinking
String
    cotton

Tape
    adding machine
    cellophane
    duct
    masking
Tape measure, metric
Thermometer, Celsius,
    non-mercury
Tongs
Trash bag

Vacuum pump
Vice grips

Washer, metal, small
Watch or clock
   with second hand
Wire, heavy
Wire cutters

Glassware

Beaker
   150-mL
   250-mL
   500-mL
Dropper

Flask, Erlenmeyer, 250-mL

Glass tubing
Graduated cylinder
   50-mL
   100-mL
   large

Jar
   glass, 1-L
   glass, gallon

Chemical Supplies

Epsom salts

Hydrochloric acid

Salt, table
Saltwater solution

Consumables

Candy, hard
Cough drop

Ice

Mineral oil

Sugar
Sugar crystal

Tap water
   cold
   hot
   room-temperature

Vegetable oil

White vinegar

Suppliers of Laboratory Materials and Equipment

Arbor Scientific
P.O. Box 2750
Ann Arbor, MI 48106
800-367-6695
arborsci.com

Carolina Biological Supply Company
2700 York Road
Burlington, NC 27215
336-584-0381
carolina.com

Central Scientific Company
11222 Melrose Ave.
Franklin Park, IL 60131
847-451-0150

Delta Biologicals
P.O. Box 26666
Tucson, AZ 58726-6666
520-745-7878
deltabio.com

Edmund Scientifics
60 Pearce Avenue
Tonawanda, NY 14150
800-728-6999
scientificsonline.com

Fisher Science Education
4500 Turnberry Drive
Hanover Park, IL 60133
800-955-1177
fishersci.com

Frey Scientific
100 Paragon Parkway
Mansfield, OH 44903
800-225-3739
freyscientific.com

Hubbard Scientific
1120 Halbleib
Chippewa Falls, WI 54729
800-289-9299
shnta.com

Lab-Aids, Inc.
17 Colt Court
Ronkonkoma, NY 11779
631-737-1133
lab-aids.com

Lab Safety Supply Inc.
401 S. Wright Rd.
Janesville, WI 53546-8729
800-356-0783
labsafety.com

Nasco–Fort Atkinson
901 Janesville Avenue
Fort Atkinson, WI 53538
800-558-9595
enasco.com

Nasco–Modesto
4825 Stoddard Road
Modesto, CA 95356-9318
800-558-9595
enasco.com

Nebraska Scientific
A Division of Cyrgus Company Inc.
3823 Leavenworth Street
Omaha, NE 68105
402-346-7214
nebraskascientific.com

Sargent-Welch Scientific Co.
911 Commerce Ct.
Buffalo Grove, IL 60089
800-727-4368
sargentwelch.com

Science Kit and Boreal Laboratories
777 East Park Drive
Tonawanda, NY 14151
800-828-7777
sciencekit.com

Ward's Natural Science Establishment Inc.
5100 West Henrietta Road
West Henrietta, NY 14586
585-359-2502
wardsci.com

Laboratory Manual

Prentice Hall
Earth Science

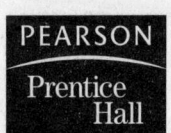

Boston, Massachusetts
Upper Saddle River, New Jersey

Laboratory Manual

Prentice Hall
Earth Science

Acknowledgments
Page **DB2 (Crystals A, B, C and E),** Jeff Scovil/Scovil Photography; **DB2 (Crystal D),** Charles D. Winters/Photo Researchers, Inc.; **DB3 (Crystal F),** Carolina Biological/Visuals Unlimited; **DB3 (Crystal G),** Breck Kent/Earth Scenes; **DB3 (Crystal H),** José Manuel Sanchis Calvete/CORBIS; **DB3 (Crystal I),** Gary Retherford/Photo Researchers, Inc.; **DB3 (Crystal J),** Jeff Scovil/Scovil Photography; **DB6–DB7,** David Sandwell, Scripps Institution of Oceanography; **DB13,** Carrie Gowran; **DB16–DB17,** Unisys Corporation.

Copyright © by Pearson Education, Inc., publishing as Pearson Prentice Hall, Boston, Massachusetts 02116. All rights reserved. Printed in the United States of America. This publication is protected by copyright, and permission should be obtained from the publisher prior to any prohibited reproduction, storage in a retrieval system, or transmission in any form or by any means, electronic, mechanical, photocopying, recording, or likewise. For information regarding permission(s), write to: Rights and Permissions Department, One Lake Street, Upper Saddle River, New Jersey 07458.

ISBN 0-13-125898-2

4 5 6 7 8 9 10   10 09 08 07 06

Contents

SI Units and Conversion Tables ix

Student Safety Manual

Science Safety Rules x

Safety Symbols xiii

Laboratory Safety Contract xiv

Student Safety Test xv

Laboratory Skills Checkup 1
Defining Elements of a Scientific Method xix

Laboratory Skills Checkup 2
Analyzing Elements of a Scientific Method xx

Laboratory Skills Checkup 3
Performing an Experiment xxi

Laboratory Skills Checkup 4
Identifying Errors xxii

Laboratory Equipment xxiii

The symbol ⚗ denotes Design Your Own Experiment Lab

Laboratory Investigations

Introduction Investigations
Investigation A *Evaluating Precision* 1

Investigation B *Measuring Volume and Temperature* 7

Chapter 1   Introduction to Earth Science
Investigation 1A *The International System of Units (SI)* 11

Investigation 1B *Using a Topographic Map to Create a Landform* 17

Chapter 2   Minerals
Investigation 2 *Crystal Systems* 23

Chapter 3   Rocks
   Investigation 3 *Classifying Rocks Using a Key* 31

Chapter 4   Earth's Resources
   Investigation 4A *Recovering Oil* 37
   Investigation 4B *Desalinization by Distillation* 41

Chapter 5   Weathering, Soil, and Mass Movements
   Investigation 5 *Some Factors That Affect Soil Erosion* 47

Chapter 6   Running Water and Groundwater
   Investigation 6A *Rivers Shape the Land* 53
   Investigation 6B *Modeling Cavern Formation* 59

Chapter 7   Glaciers, Deserts, and Wind
   Investigation 7 *Continental Glaciers Change Earth's Topography* 65

Chapter 8   Earthquakes and Earth's Interior
   Investigation 8A *Modeling Liquefaction* 69
   Investigation 8B *Designing and Building a Simple Seismograph* 73

Chapter 9   Plate Tectonics
   Investigation 9 *Modeling a Plate Boundary* 79

Chapter 11   Mountain Building
   Investigation 11 *Interpreting a Geologic Map* 85

Chapter 12   Geologic Time
   Investigation 12 *Modeling Radioactive Decay* 89

Chapter 13   Earth's History
   Investigation 13 *Determining Geologic Ages* 93

Chapter 14   The Ocean Floor
   Investigation 14 *Modeling the Ocean Floor* 97

Chapter 16   The Dynamic Ocean
   Investigation 16 *Shoreline Features* 101

Chapter 17 The Atmosphere: Structure and Temperature
Investigation 17A *Determining How Temperature Changes with Altitude* ... 107

Investigation 17B *Investigating Factors That Control Temperature* 111

Chapter 18 Moisture, Clouds, and Precipitation
Investigation 18 *Recipe for a Cloud* 115

Chapter 19 Air Pressure and Wind
Investigation 19 *Analyzing Pressure Systems* 119

Chapter 20 Weather Patterns and Severe Storms
Investigation 20A *Analyzing Severe Weather Data* 123

Investigation 20B *Interpreting Weather Diagrams* 129

Investigation 20C *Creating a Weather Station* 133

Chapter 21 Climate
Investigation 21 *Modeling the Greenhouse Effect* 137

Chapter 22 Origin of Modern Astronomy
Investigation 22 *Measuring the Angle of the Sun at Noon* 141

Chapter 23 Touring Our Solar System
Investigation 23 *Exploring Orbits* 145

Chapter 24 Studying the Sun
Investigation 24 *Measuring the Diameter of the Sun* 151

Chapter 25 Beyond Our Solar System
Investigation 25 *Modeling the Rotation of Neutron Stars* 157

Student Edition Lab Worksheets

Chapter 1 Introduction to Earth Science
Exploration Lab
Determining Longitude and Latitude 161

Chapter 2 Minerals
Exploration Lab
Mineral Identification 165

Chapter 3 Rocks
Exploration Lab
Rock Identification 169

Chapter 4 Earth's Resources
Application Lab
Finding the Product That Best Conserves Resources 171

Chapter 5 Weathering, Soil, and Mass Movements
Exploration Lab
Effect of Temperature on Chemical Weathering 175

Chapter 6 Running Water and Groundwater
Exploration Lab
Investigating the Permeability of Soils 179

Chapter 7 Glaciers, Deserts, and Wind
Exploration Lab
Interpreting a Glacial Landscape 181

Chapter 8 Earthquakes and Earth's Interior
Exploration Lab
Locating an Earthquake 183

Chapter 9 Plate Tectonics
Exploration Lab
Paleomagnetism and the Ocean Floor 185

Chapter 10 Volcanoes and Other Igneous Activity
Exploration Lab
Melting Temperatures of Rocks 189

Chapter 11 Mountain Building
Exploration Lab
Investigating Anticlines and Synclines 191

Chapter 12 Geologic Time
Exploration Lab
Fossil Occurrence and the Age of Rocks 195

Chapter 13 Earth's History
Application Lab
Modeling the Geologic Time Scale 197

Chapter 14 The Ocean Floor
Exploration Lab
Modeling Seafloor Depth Transects 199

Chapter 15 Ocean Water and Ocean Life
Exploration Lab
How Does Temperature Affect Water Density? 203

Chapter 16 The Dynamic Ocean
Exploration Lab
Graphing Tidal Cycles 207

Chapter 17 The Atmosphere: Structure and Temperature
Exploration Lab
Heating Land and Water 209

Chapter 18 Moisture, Clouds, and Precipitation
Exploration Lab
Measuring Humidity 211

Chapter 19 Air Pressure and Wind
Exploration Lab
Observing Wind Patterns 215

Chapter 20 Weather Patterns and Severe Storms
Exploration Lab
Middle-Latitude Cyclones 217

Chapter 21 Climate
Exploration Lab
Human Impact on Climate and Weather 219

Chapter 22 Origin of Modern Astronomy
Exploration Lab
Modeling Synodic and Sidereal Months 223

Chapter 23 Touring Our Solar System
Exploration Lab
Modeling the Solar System 225

Chapter 24 Studying the Sun
Exploration Lab
Tracking Sunspots 227

Chapter 25 Beyond Our Solar System
Exploration Lab
Observing Stars 229

DataBank

Resource 1	Map Symbols	DB1
Resource 2	Identifying Crystal Systems	DB2
Resource 3	Earth's Tectonic Plates	DB4
Resource 4	Global Bathymetry from Altimetry Data	DB6
Resource 5	Ridge Fracture Zone	DB7
Resource 6	Geologic Map of Devil's Fence, Montana	DB8
Resource 7	Map Key for Geologic Map	DB9
Resource 8	Topographic Map of Campti, Louisiana	DB10
Resource 9	Topographic Map of Whitewater, Wisconsin	DB11
Resource 10	The Geologic Time Scale	DB12
Resource 11	Key to Index Fossils	DB13
Resource 12	Atmospheric Temperature Curve	DB14
Resource 13	Dew-Point Temperature Table	DB15
Resource 14	Temperature Contour Plots	DB16
Resource 15	Temperature Change and Heat Index Plots	DB17
Resource 16	Some Common Minerals and Their Properties	DB18
Resource 17	Classification of Rocks	DB20
Resource 18	Topography of the Ocean Floor	DB22
Resource 19	Star Charts	DB24
Resource 20	Landforms of the Conterminous United States	DB26
Resource 21	Circulation on a Rotating Earth/ Hertzsprung-Russel Diagram	DB28
Resource 22	Middle-Latitude Cyclone Model	DB29
Resource 23	Topographic Map of a Glacial Landscape	DB30

SI Units and Conversion Table

COMMON SI UNITS

Measurement	Unit	Symbol	Equivalents
Length	1 millimeter	mm	1,000 micrometers (µm)
	1 centimeter	cm	10 millimeters (mm)
	1 meter	m	100 centimeters (cm)
	1 kilometer	km	1,000 meters (m)
Area	1 square meter	m^2	10,000 square centimeters (cm^2)
	1 square kilometer	km^2	1,000,000 square meters (m^2)
Volume	1 milliliter	mL	1 cubic centimeter (cm^3 or cc)
	1 liter	L	1,000 milliliters (mL)
Mass	1 gram	g	1,000 milligrams (mg)
	1 kilogram	kg	1,000 grams (g)
	1 ton	t	1,000 kilograms (kg) = 1 metric ton
Time	1 second	s	
Temperature	1 Kelvin	K	1 degree Celsius (°C)

METRIC CONVERSION TABLES

When You Know	Multiply by	To Find			
		When You Know	Multiply by	To Find	
inches	2.54	centimeters	0.394	inches	
feet	0.3048	meters	3.281	feet	
yards	0.914	meters	1.0936	yards	
miles	1.609	kilometers	0.62	miles	
square inches	6.45	square centimeters	0.155	square inches	
square feet	0.093	square meters	10.76	square feet	
square yards	0.836	square meters	1.196	square yards	
acres	0.405	hectares	2.471	acres	
square miles	2.59	square kilometers	0.386	square miles	
cubic inches	16.387	cubic centimeters	0.061	cubic inches	
cubic feet	0.028	cubic meters	35.315	cubic feet	
cubic yards	0.765	cubic meters	1.31	cubic yards	
fluid ounces	29.57	milliliters	0.0338	fluid ounces	
quarts	0.946	liters	1.057	quarts	
gallons	3.785	liters	0.264	gallons	
ounces	28.35	grams	0.0353	ounces	
pounds	0.4536	kilograms	2.2046	pounds	
tons	0.907	metric tons	1.102	tons	

When You Know		
Fahrenheit	subtract 32; then divide by 1.8	to find Celsius
Celsius	multiply by 1.8; then add 32	to find Fahrenheit
Celsius	add 273	to find Kelvin

Science Safety Rules

To prepare yourself to work safely in the laboratory, read over the following safety rules. Then read them a second time. Make sure you understand and follow each rule. Ask your teacher to explain any rules you do not understand.

Dress Code
1. To protect yourself from injuring your eyes, wear safety goggles whenever you work with chemicals, flames, glassware, or any substance that might get into your eyes. If you wear contact lenses, notify your teacher.
2. Wear an apron or coat whenever you work with corrosive chemicals or substances that can stain.
3. Tie back long hair to keep it away from any chemicals, flames, or equipment.
4. Remove or tie back any article of clothing or jewelry that can hang down and touch chemicals, flames, or equipment. Roll up or secure long sleeves.
5. Never wear open shoes or sandals.

General Precautions
6. Read all directions for an experiment several times before beginning the activity. Carefully follow all written and oral instructions. If you are in doubt about any part of the experiment, ask your teacher for assistance.
7. Never perform activities that are not assigned or authorized by your teacher. Obtain permission before "experimenting" on your own. Never handle any equipment unless you have specific permission.
8. Never perform lab activities without direct supervision.
9. Never eat or drink in the laboratory.
10. Keep work areas clean and tidy at all times. Bring only notebooks and lab manuals or written lab procedures to the work area. All other items, such as purses and backpacks, should be left in a designated area.
11. Do not engage in horseplay.

First Aid
12. Always report all accidents or injuries to your teacher, no matter how minor. Notify your teacher immediately about any fires.
13. Learn what to do in case of specific accidents, such as getting acid in your eyes or on your skin. (Rinse acids from your body with plenty of water.)
14. Be aware of the location of the first-aid kit, but do not use it unless instructed by your teacher. In case of injury, your teacher should administer first aid. Your teacher may also send you to the school nurse or call a physician.
15. Know the location of the emergency equipment such as fire extinguisher and fire blanket.
16. Know the location of the nearest telephone and whom to contact in an emergency.

Heating and Fire Safety
17. Never use a heat source, such as a candle, burner, or hot plate, without wearing safety goggles.
18. Never heat anything unless instructed to do so. A chemical that is harmless when cool may be dangerous when heated.
19. Keep all combustible materials away from flames. Never use a flame or spark near a combustible chemical.
20. Never reach across a flame.
21. Before using a laboratory burner, make sure you know proper procedures for lighting and adjusting the burner, as demonstrated by your teacher. Do not touch the burner. It may be hot. Never

Earth Science Lab Manual

leave a lighted burner unattended. Turn off the burner when not in use.
22. Chemicals can splash or boil out of a heated test tube. When heating a substance in a test tube, make sure that the mouth of the tube is not pointed at you or anyone else.
23. Never heat a liquid in a closed container. The expanding gases produced may shatter the container.
24. Before picking up a container that has been heated, first hold the back of your hand near it. If you can feel heat on the back of your hand, the container is too hot to handle. Use an oven mitt to pick up a container that has been heated.

Using Chemicals Safely
25. Never mix chemicals "for the fun of it." You might produce a dangerous, possibly explosive substance.
26. Never put your face near the mouth of a container that holds chemicals. Many chemicals are poisonous. Never touch, taste, or smell a chemical unless you are instructed by your teacher to do so.
27. Use only those chemicals needed in the activity. Read and double-check labels on supply bottles before removing any chemicals. Take only as much as you need. Keep all containers closed when chemicals are not being used.
28. Dispose of all chemicals as instructed by your teacher. To avoid contamination, never return chemicals to their original containers. Never pour untreated chemicals or other substances into the sink or trash containers.
29. Be extra careful when working with acids or bases. Pour all chemicals over the sink or a container, not over your work surface.
30. If you are instructed to test for odors, use a wafting motion to direct the odors to your nose. Do not inhale the fumes directly from the container.
31. When mixing an acid and water, always pour the water into the container first then add the acid to the water. Never pour water into an acid.
32. Take extreme care not to spill any material in the laboratory. Wash chemical spills and splashes immediately with plenty of water. Immediately begin rinsing with water any acids that get on your skin or clothing, and notify your teacher of any acid spill at the same time.

Using Glassware Safely
33. Never force glass tubing or a thermometer into a rubber stopper or rubber tubing. Have your teacher insert the glass tubing or thermometer if required for an activity.
34. If you are using a laboratory burner, use a wire screen to protect glassware from any flame. Never heat glassware that is not thoroughly dry on the outside.
35. Keep in mind that hot glassware looks cool. Never pick up glassware without first checking to see if it is hot. Use an oven mitt. See rule 24.
36. Never use broken or chipped glassware. If glassware breaks, notify your teacher and dispose of the glassware in the proper broken-glassware container.
37. Never eat or drink from glassware.
38. Thoroughly clean glassware before putting it away.

Using Sharp Instruments
39. Handle scalpels or other sharp instruments with extreme care. Never cut material toward you; cut away from you.
40. Immediately notify your teacher if you cut your skin while working in the laboratory.

Field Safety

41. When leaving the classroom or in the field, do not disrupt the activities of others. Do not leave your group unless you notify a teacher first.
42. Your teacher will instruct you as to how to conduct your research or experiment outside the classroom.
43. Never touch any animals or plants that you encounter in the field unless your teacher instructs you in the proper handling of that species.
44. Clean your hands thoroughly after handling anything in the field.

End-of-Experiment Rules

45. After an experiment has been completed, turn off all burners or hot plates. If you used a gas burner, check that the gas-line valve to the burner is off. Unplug hot plates.
46. Turn off and unplug any other electrical equipment that you used.
47. Clean up your work area and return all equipment to its proper place.
48. Dispose of waste materials as instructed by your teacher.
49. Wash your hands after every experiment.

Safety Symbols

These symbols alert you to possible dangers in the laboratory and remind you to work carefully.

General Safety Awareness You may see this symbol when none of the symbols described below appears. In this case, follow the specific instructions provided. You may also see this symbol when you are asked to develop your own procedure in a lab. Have the teacher approve your plan before you go further.

Physical Safety When an experiment involves physical activity, take precautions to avoid injuring yourself or others. Follow instructions from the teacher. Alert the teacher if there is any reason you should not participate in the activity.

Safety Goggles Always wear safety goggles to protect your eyes in any activity involving chemicals, flames or heating, or the possibility of broken glassware.

Lab Apron Wear a laboratory apron to protect your skin and clothing from damage.

Plastic Gloves Wear disposable plastic gloves to protect yourself from chemicals or organisms that could be harmful. Keep your hands away from your face. Dispose of the gloves according to your teacher's instructions at the end of the activity.

Heating Use a clamp or tongs to pick up hot glassware. Do not touch hot objects with your bare hands.

Heat-Resistant Gloves Use an oven mitt or other hand protection when handling hot materials. Hot plates, hot glassware, or hot water can cause burns. Do not touch hot objects with your bare hands.

Flames You may be working with flames from a lab burner, candle, or matches. Tie back loose hair and clothing. Follow instructions from the teacher about lighting and extinguishing flames.

No Flames Flammable materials may be present. Make sure there are no flames, sparks, or other exposed heat sources present.

Electric Shock Avoid the possibility of electric shock. Never use electrical equipment around water, or when the equipment is wet or your hands are wet. Be sure cords are untangled and cannot trip anyone. Disconnect the equipment when it is not in use.

Breakage You are working with materials that may be breakable, such as glass containers, glass tubing, thermometers, or funnels. Handle breakable materials with care. Do not touch broken glassware.

Corrosive Chemical You are working with an acid or another corrosive chemical. Avoid getting it on your skin or clothing, or in your eyes. Do not inhale the vapors. Wash your hands when you are finished with the activity.

Poison Do not let any poisonous chemical come in contact with your skin, and do not inhale its vapors. Wash your hands when you are finished with the activity.

Fumes When poisonous or unpleasant vapors may be involved, work in a ventilated area. Avoid inhaling vapors directly. Only test an odor when directed to do so by the teacher, and use a wafting motion to direct the vapor toward your nose.

Sharp Object Pointed-tip scissors, scalpels, knives, needles, pins, or tacks are sharp. They can cut or puncture your skin. Always direct a sharp edge or point away from yourself and others. Use sharp instruments only as instructed.

Disposal Chemicals and other laboratory materials used in the activity must be disposed of safely. Follow the instructions from the teacher.

Hand Washing Wash your hands thoroughly when finished with the activity. Use antibacterial soap and warm water. Lather both sides of your hands and between your fingers. Rinse well.

LABORATORY SAFETY CONTRACT

I, _____ , have read the Science Safety Rules and Safety Symbols sections on pages x–xiii of this manual, understand their contents completely, and agree to demonstrate compliance with all safety rules and guidelines that have been established in each of the following categories:

(please check)

☐ Dress Code   ☐ Using Glassware Safely

☐ General Precautions   ☐ Using Sharp Instruments

☐ First Aid   ☐ Field Safety

☐ Heating and Fire Safety ☐ End-of-Experiment Rules

☐ Using Chemicals Safely

Signature _____

Date _____

Name _____ Class _____ Date _____

STUDENT SAFETY TEST

Recognizing Laboratory Safety

TIME REQUIRED: 40 minutes

Pre-Lab Discussion

An important part of your study of science will be working in a laboratory. In the laboratory, you and your classmates will learn about the natural world by conducting experiments. Working directly with household objects, laboratory equipment, and even living things will help you to better understand the concepts you read about in your textbook or in class.

Most of the laboratory work you will do is quite safe. However, some laboratory equipment, chemicals, and specimens can be dangerous if handled improperly. Laboratory accidents do not just happen. They are caused by carelessness, improper handling of equipment, or inappropriate behavior.

In this investigation, you will learn how to prevent accidents and thus work safely in a laboratory. You will review some safety guidelines and become acquainted with the location and proper use of safety equipment in your classroom laboratory.

Problem

What are the proper practices for working safely in a science laboratory?

Materials (per group)

Be sure to show the location of all safety equipment in your laboratory. Also give instructions on its proper use. Guidelines pertaining to the use of special equipment, fire-drill procedures, or penalties for misbehavior in the lab might also be discussed at this time.

Science textbook
Laboratory safety equipment (for demonstration)

Procedure

Part A: Reviewing Laboratory Safety Rules and Symbols

1. Carefully read the list of laboratory safety rules listed on pages x – xii of this lab manual.
2. Special symbols are used throughout this lab book to call attention to investigations that require extra caution. Use pages xii and xiii as a reference to describe what each symbol means in numbers l through 7 of Observations.

Part B: Location of Safety Equipment in Your Science Laboratory

1. The teacher will point out the location of the safety equipment in your classroom laboratory. Pay special attention to instructions for using such equipment as fire extinguishers, eyewash fountains, fire blankets, safety showers, and items in first-aid kits. Use the space provided in Part B under Observations to list the location of all safety equipment in your laboratory.

Earth Science Lab Manual ▪ XV

Name _____ Class _____ Date _____

RECOGNIZING LABORATORY SAFETY (continued)

Observations

Part A

1. Student is working with materials that can easily be broken, such as glass containers or thermometers. They should be handled carefully, and broken glassware should not be touched.

2. Student is working with a flame and should tie back loose hair and clothing.

3. Student should use oven mitts or other hand protection to avoid burning hands.

4. Student is working with poisonous chemicals and should not let the chemical touch the skin or inhale its vapors. Student should wash hands after the lab.

5. Student is performing an experiment in which the eyes and face should be protected by safety goggles.

6. Student is working with a sharp instrument and should direct the sharp edge or point away from himself or herself and others.

7. Student is using electricity in the laboratory and should avoid the possibility of electric shock. Electrical equipment should not be used around water, cords should not be tangled, and equipment should be disconnected when not in use.

Name _____ Class _____ Date _____

RECOGNIZING LABORATORY SAFETY (continued)

Part B

Student responses will depend on the specific safety features of your classroom laboratory. Locations might include such directions as above the sink, to the right of the goggles case, near the door, and so on.

Analyze and Conclude

Look at each of the following drawings and explain why the laboratory activities pictured are unsafe.

1. Safety goggles should always be worn whenever a person is working with chemicals, lab burners, or any substance that might get into the eye.

2. When diluting an acid, pour the acid into water. Never pour water into the acid. Also, safety goggles and a lab apron should be worn when working with chemicals.

3. Never heat a liquid in a closed container. The expanding gases produced may shatter the container.

Earth Science Lab Manual ▪ xvii

Name _____ Class _____ Date _____

RECOGNIZING LABORATORY SAFETY (continued)

Critical Thinking and Applications

In each of the following situations, write yes if the proper safety procedures are being followed and no if they are not. Then give a reason for your answer.

1. Gina is thirsty. She rinses a beaker with water, refills it with water, and takes a drink.
   No; you should never drink from laboratory glassware. The last substance in it may have been poisonous and traces of the poison may remain.

2. Bram notices that the electrical cord on his microscope is frayed near the plug. He takes the microscope to his teacher and asks for permission to use another one.
   Yes; electrical appliances with frayed cords or broken insulation may present a hazard and should not be used.

3. The printed directions in the lab book tell a student to pour a small amount of hydrochloric acid into a beaker. Jamal puts on safety goggles before pouring the acid into the beaker.
   Yes; safety goggles should always be worn when working with dangerous chemicals.

4. It is rather warm in the laboratory during a late spring day. Anna slips off her shoes and walks barefoot to the sink to clean her glassware.
   No; shoes should always be kept on while working in the laboratory in case glassware breaks or chemicals are spilled onto the floor.

5. While washing glassware, Mike splashes some water on Evon. To get even, Evon splashes him back.
   No; misbehaving is never acceptable in a laboratory.

6. During an experiment, Lindsey decides to mix two chemicals that the lab procedure does not say to mix, because she is curious about what will happen.
   No; never mix chemicals unless directed to do so. The mixing might produce an explosive substance.

Name _____ Class _____ Date _____

Laboratory Skills Checkup 1

Defining Elements of a Scientific Method

Laboratory activities and experiments involve the use of the scientific method. Listed in the left column are the names of parts of this method. The right column contains definitions. Next to each word in the left column, write the letter of the definition that best matches that word.

__A__ 1. Hypothesis A. Prediction about the outcome of an experiment

__E__ 2. Manipulated Variable B. What you measure or observe to obtain your results

__B__ 3. Responding Variable C. Measurements and other observations

__G__ 4. Controlling Variables D. Statement that sums up what you learn from an experiment

__F__ 5. Observation E. Factor that is changed in an experiment

__C__ 6. Data F. What the person performing the activity sees, hears, feels, smells, or tastes

__D__ 7. Conclusion G. Keeping all variables the same except the manipulated variable

Name _____ Class _____ Date _____

Laboratory Skills Checkup 2

Analyzing Elements of a Scientific Method

Read the following statements and then answer the questions.

1. You and your friend are walking along a beach in Maine on January 15, at 8:00 am.
2. You notice a thermometer on a nearby building that reads −1°C.
3. You also notice that there is snow on the roof of the building and icicles hanging from the roof.
4. You further notice a pool of sea water in the sand near the ocean.
5. Your friend looks at the icicles and the pool and says, "How come the water on the roof is frozen and the sea water is not?"
6. You answer, "I think that the salt in the sea water keeps it from freezing at −1°C."
7. You go on to say, "And I think under the same conditions, the same thing will happen tomorrow."
8. Your friend asks, "How can you be sure?" You answer, "I'm going to get some fresh water and some salt water and expose them to a temperature of −1°C and see what happens."

Questions

A. In which statement is a prediction made?

7

B. Which statement states a problem?

5

C. In which statement is an experiment described?

8

D. Which statement contains a hypothesis?

6

E. Which statements contain data?

1, 2, 3, 4

F. Which statements describe observations?

2, 3, 4

Name _____ Class _____ Date _____

Laboratory Skills Checkup 3

Performing an Experiment

Read the following statements and then answer the questions.

1. A scientist wants to find out why sea water freezes at a lower temperature than fresh water.
2. The scientist goes to the library and reads a number of articles about the physical properties of solutions.
3. The scientist also reads about the composition of sea water.
4. The scientist travels to a nearby beach and observes the conditions there. The scientist notes the taste of the sea water and other factors such as waves, wind, air pressure, temperature, and humidity.
5. After considering all this information, the scientist sits at a desk and writes, "If sea water has salt in it, it will freeze at a lower temperature than fresh water."
6. The scientist goes back to the laboratory and does the following:
   a. Fills each of two beakers with 1 liter of fresh water.
   b. Dissolves 35 grams of table salt in one of the beakers.
   c. Places both beakers in a freezer at a temperature of −1°C.
   d. Leaves the beakers in the freezer for 24 hours.
7. After 24 hours, the scientist examines both beakers and finds the fresh water to be frozen. The salt water is still liquid.
8. The scientist writes in a notebook, "It appears that salt water freezes at a lower temperature than fresh water does."
9. The scientist continues, "I suggest that the reason sea water freezes at a lower temperature is that sea water contains dissolved salts, while fresh water does not."

Questions

A. Which statement(s) contain conclusions? _____8, 9_____

B. Which statement(s) contains a hypothesis? _____5_____

C. Which statement(s) contain observations? _____4, 7_____

D. Which statement(s) describe an experiment? _____6 a–d_____

E. In which statement is the problem described? _____1_____

F. Which statement(s) contain data? _____4, 6 a–d, 7_____

G. Which is the manipulated variable in the experiment? ___the amount of salt in water___

H. What is the responding variable in the experiment? ___the temperature at which water freezes___

Name _____ Class _____ Date _____

Laboratory Skills Checkup 4

Identifying Errors

Read the following paragraph and then answer the questions.

Andrew arrived at school and went directly to his earth science class. He took off his cap and coat and sat down at his desk. His teacher gave him a large rock and asked him to find its density. Realizing that the rock was too large to work with, Andrew got a hammer from the supply cabinet and hit the rock several times until he broke off a chip small enough to work with. He partly filled a graduated cylinder with water and suspended the rock in the water. The water level rose 2 cm. Andrew committed this measurement to memory. He next weighed the rock on a balance. The rock weighed 4 oz. Andrew then calculated the density of the rock as follows: He divided 2 cm by 4 oz. He then reported to his teacher that the density of the rock was 0.5 cm/oz.

Questions

1. What safety rule(s) did Andrew break?

   He didn't put on his safety goggles. Also, he didn't obtain permission from his teacher before obtaining the hammer and breaking the rock.

2. What mistake did Andrew make using measurement units?

   He used linear units (centimeters) instead of volumetric units (milliliters).

3. What should Andrew have done with his data rather than commit them to memory?

   He should have kept a written record.

4. What is wrong with the statement "He next weighed the rock on a balance"?

   A balance is used to determine mass, not weight.

5. Why is "4 oz" an inappropriate measurement in a science experiment?

   Metric units (grams) should be used.

6. What mistake did Andrew make in calculating density?

   Density is expressed in mass per unit volume (g/mL), not length per unit weight.

Some Common Laboratory Equipment

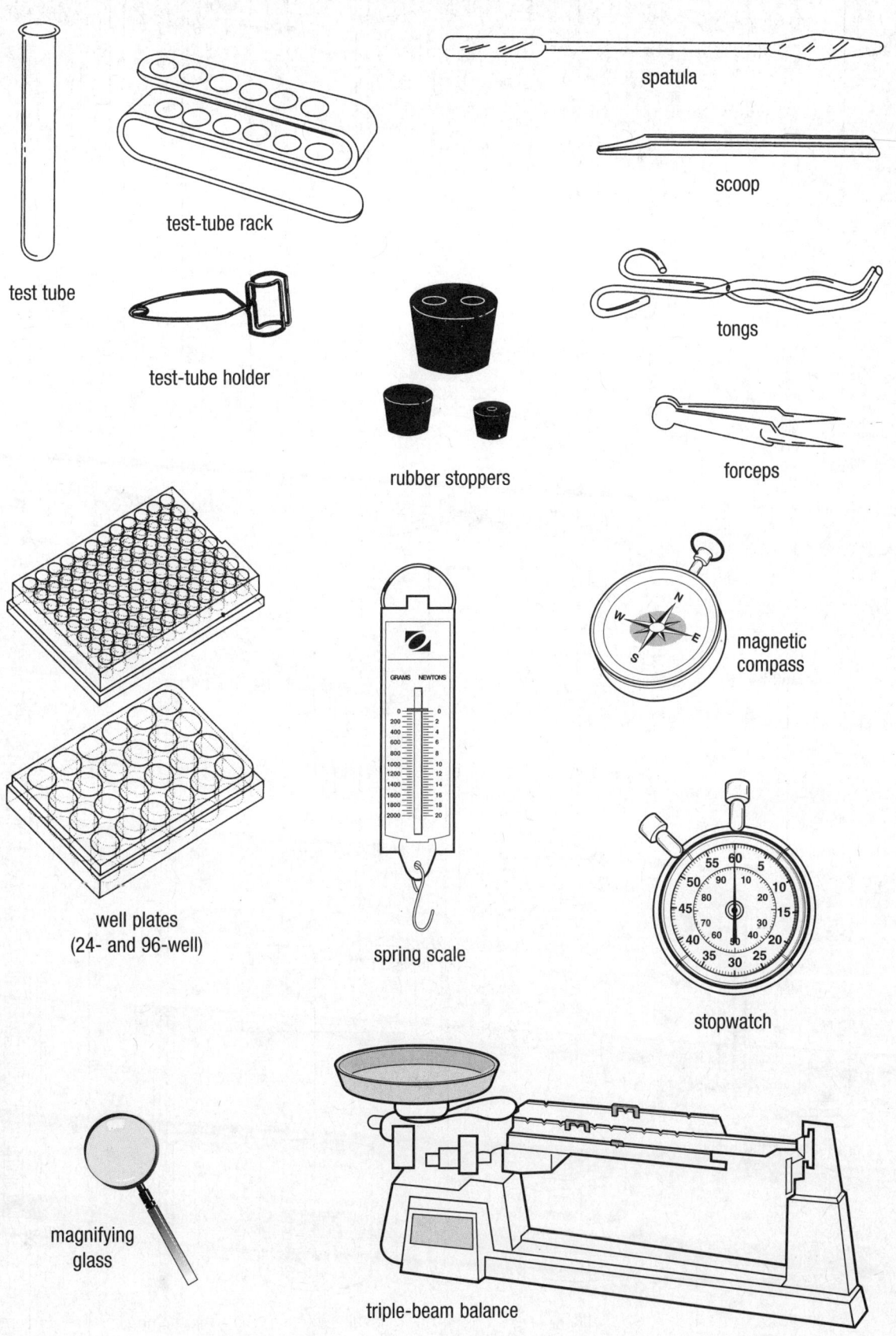

SOME COMMON LABORATORY EQUIPMENT (continued)

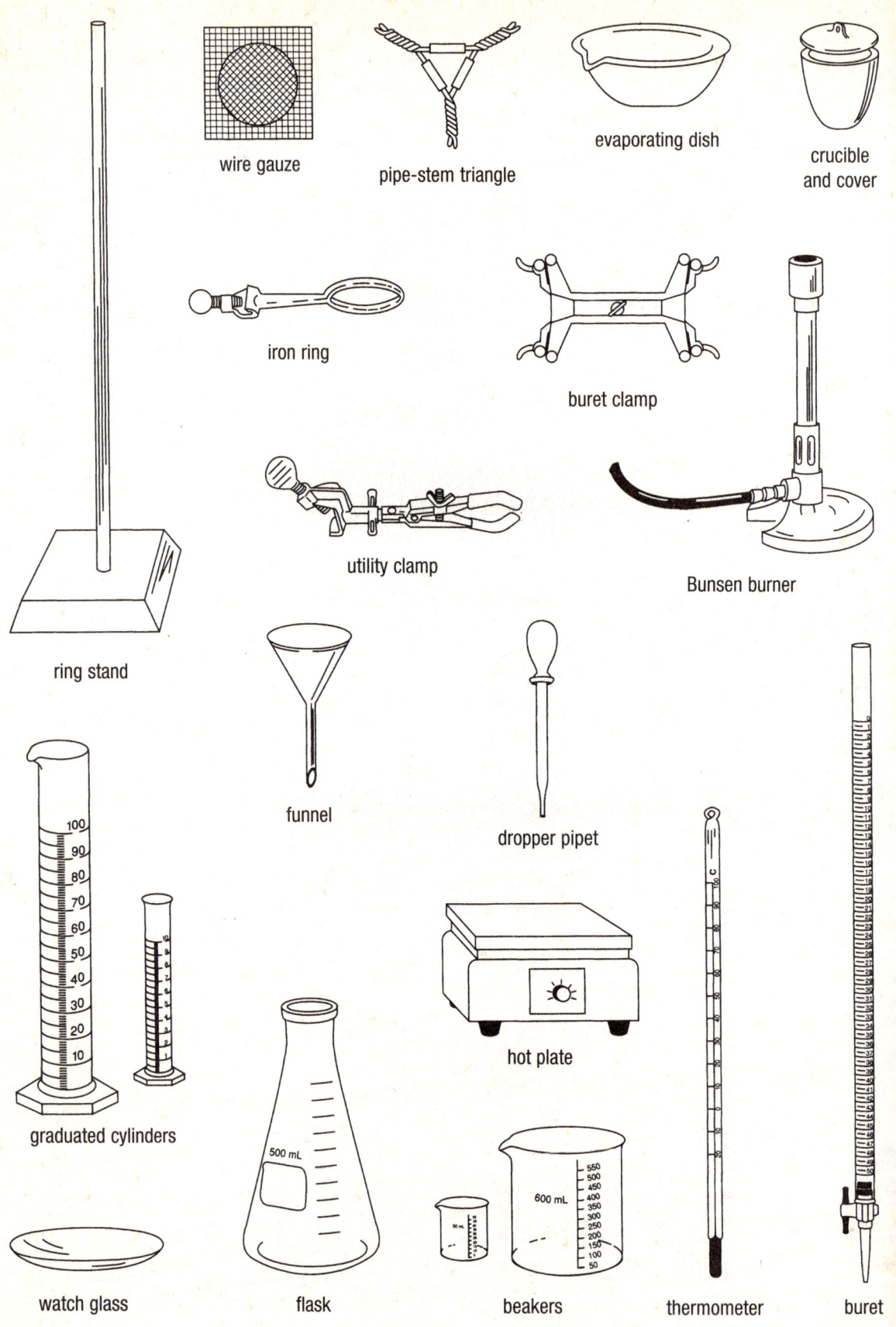

Name _____ Class _____ Date _____

Science Skills   Introduction Investigation A

Evaluating Precision
SKILLS FOCUS: Measuring, Evaluating
TIME REQUIRED: 30 minutes

Introduction

When an object is measured more than once, the measurements may vary. The closeness of a set of measured values to each other is called **precision.** Many people confuse precision with accuracy. **Accuracy** is a measure of how close the values are to the actual value. A set of values can be in close agreement, or precise, without being accurate.

For example, suppose you repeatedly measure the mass of a 4.00-g mineral sample by using a balance that reads too low by 3.00 g every time. You might get nearly identical readings—for example, 1.00 g, 1.01 g, and 0.99 g. These readings are quite precise because they are close together. However, they differ from the actual value by a large amount. Therefore, the measurements are very inaccurate.

In this investigation, you will make several measurements of length, temperature, and volume. Then, you will evaluate the precision of your measurements by comparing them to measurements made by your classmates.

Problem

How can you determine the precision and accuracy of measurements?

Pre-Lab Discussion

Read the entire investigation. Then work with a partner to answer the following questions.

1. **Applying Concepts** Use the example of a series of repeated length measurements to explain the meaning of precision.

   The precision of a series of repeated measurements describes how closely the individual measurements agree. For example, if a meter stick is used to measure the length of a desk three times, the measurements 75 cm, 74 cm, and 76 cm are more precise than measurements of 70 cm, 75 cm, and 80 cm.

2. **Inferring** What information would you need to determine the accuracy of a measurement?

   You would need a reliable measurement to which you could compare your measurement.

Name _____ Class _____ Date _____

3. Drawing Conclusions In this investigation, you will compare measurements that you make to measurements that your classmates make. Will you do this to determine the accuracy or the precision of your measurements?

Comparing measurements made by different students can determine precision, but not accuracy.

4. Designing Experiments Identify the manipulated, responding, and controlled variables in this investigation.

   a. Manipulated variable

   teams making the measurements

   b. Responding variable

   precision of measurements

   c. Controlled variables

   devices for measuring length, temperature, and volume; the objects whose properties are being measured

5. Analyzing Data Two students measure the mass of a wooden disk, using the same balance. The first student repeats the weighing three times and obtains mass readings of 47 g, 52 g, and 51 g. The second student obtains mass readings of 45 g, 55 g, and 50 g. Explain which set of measurements is more precise. Can you tell if the measurements are accurate? Why or why not?

The first student's measurements are more precise because the values are closer together. Although both sets of measurements suggest that the disk has a mass of nearly 50 g, this may not be true. The balance may not be properly calibrated, so it could be providing consistently inaccurate measurements.

Name _____ Class _____ Date _____

Materials *(per group)*
meter stick
Celsius thermometer *Provide only non-mercury thermometers.*
500-mL beaker filled with room-temperature water *Fill a 500-mL beaker with water and allow it to sit at room temperature for at least 1 hour before class.*
10 pennies
50-mL graduated cylinder

Safety
Put on safety goggles and a lab apron. Be careful to avoid breakage when working with glassware. Note all safety alert symbols next to the steps in the Procedure and review the meaning of each symbol by referring to the Safety Symbols on page xiii. *You may wish to start with one lab group measuring length, a second group measuring temperature, and a third group measuring volume, and then rotate the groups to minimize waiting time between measurements. In any case, make sure all students within a group use the same measuring devices.*

Procedure
1. You and your partner make up a team. Your team and two other teams will make up a group of six. Your teacher will tell you and your partner whether you are Team A, B, or C of your group. The three teams in your group will measure the same objects separately. You will not share your measurements with the other teams in your group until you complete the procedure.

2. Working with your partner, use the meter stick to measure the length of a desk indicated by your teacher. Measure as carefully as possible, to the nearest millimeter. Record the length of the desk in the Data Table. (*Hint:* Do not reveal the measurements you make to the other teams in your group. They must make the same measurements and must not be influenced by your results.)

3. Use the thermometer to measure the temperature of the beaker of room-temperature water. **CAUTION:** *Do not let the thermometer touch the beaker.* Record this measurement in the Data Table.

4. Place 25 mL of tap water in the graduated cylinder. Measure the volume of the water. Record this volume in the Data Table to the nearest 0.1 mL. (*Hint:* Remember to read the volume at the bottom of the meniscus.) *You may need to show students how to read a liquid volume at the bottom of the meniscus.*

5. Add the 10 pennies to the graduated cylinder. Read the volume of the water and pennies. Record the volume the nearest 0.1 mL in the Data Table.

6. Subtract the volume of the water from the volume of the water and pennies. The result is a measurement of the volume of the pennies. Record this value in the Data Table.

7. After all three teams in your group have finished measuring the same objects for length, temperature, and volume, share your results with the other two teams. Record their measurements in the Data Table.

Name _____ Class _____ Date _____

Observations Sample data are shown.

DATA TABLE

Measurement	Team A	Team B	Team C
Length of desk (mm)	400.0–450.0		
Temperature of water (°C)	18.0–21.0		
Volume of water (mL)	24.5–25.5		
Volume of water and pennies (mL)	27.5–29.0		
Volume of pennies (mL)	3.0–3.5		

Analysis and Conclusions

1. **Calculating** Average the three length measurements you compared by adding them together and dividing the result by 3. Find the range of values by calculating the difference between the largest and smallest values. Record the results of your calculations in the space below.

   Calculations should be correctly carried out. Actual results will depend on student data.

   a. Average of length measurements (mm)
   _____

   b. Range of length measurements (mm)
   _____

Earth Science Lab Manual ▪ 4

Name _____ Class _____ Date _____

2. Making Generalizations Would it be correct to use the range of values you calculated in Question 1 to describe the precision of the measurements? The accuracy of the measurements? Explain your answer.

The range of values can be used to describe the precision of the measurements. The narrower the range is, the more precise the measurements are. The range cannot be used to describe the accuracy of the measurements, which depends on the difference between the measurements and the actual value.

3. Analyzing Data Which of the three sets of measurements had the least spread among the measurements? Suggest reasons for the precision of these measurements.

Answers may vary for each group. Reasons for good precision in a set of measurements depend on the smaller units with which the measuring instrument can be read, the consistency of technique in making and reading measurements, and that the conditions under which the measurements are made are kept the same.

4. Applying Concepts Figure 1 shows the results of three people's attempts to shoot as many bull's-eyes as possible. Below Figure 1, label each of the results as *accurate* or *not accurate,* and as *precise* or *not precise*.

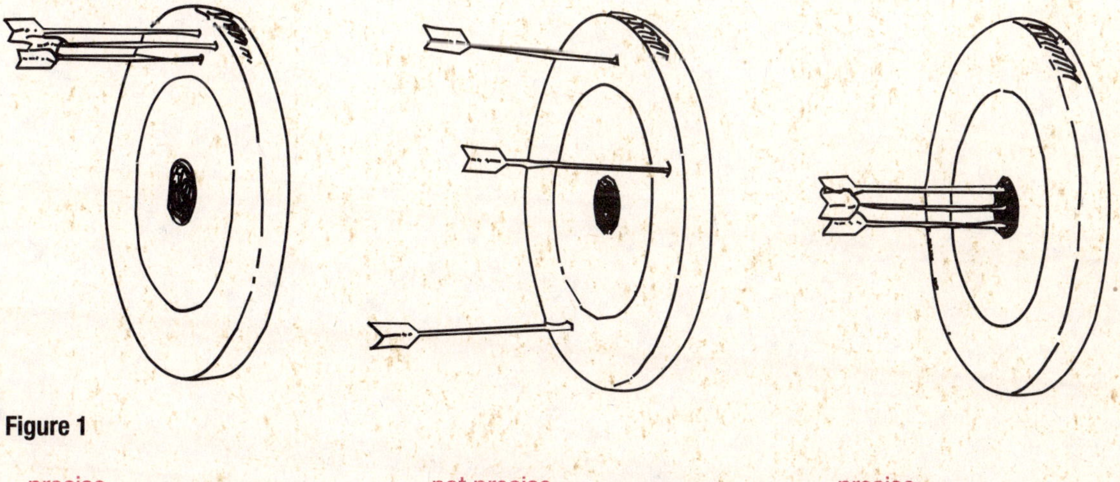

Figure 1

precise   not precise   precise
not accurate   not accurate   accurate

Name _____ Class _____ Date _____

5. Evaluating and Revising Discuss the reasons for the differences among the teams' measurements with the members of your group. Describe these reasons and explain how the measurements could be made more precise.

Students may cite skill and care of measurement and the limited precision of the measuring instruments as factors that limited the precision of the measurements. Students may suggest having one person make all the measurements or using more precise instruments.

Go Further

Design an experiment to compare the precision of two or more measuring instruments. Is the precision of each instrument the same throughout its range of measurements? Write a procedure you would follow to answer these questions. After your teacher approves your procedure, carry out the experiment and report your results.

Measuring instruments differ in precision. For example, a micrometer is more precise than a ruler. The precision of a well-made measuring instrument is nearly the same throughout its range of measurements except for a decline in precision for values near zero. At very low values, the precision declines as the size of the measurement approaches the sensitivity of the instrument, which is the smallest nonzero value it can measure.

Name _____ Class _____ Date _____

Science Skills Introduction Investigation B

Measuring Volume and Temperature

Introduction
SKILLS FOCUS: Measuring **TIME REQUIRED:** 40 minutes

The amount of space an object takes up is called its volume. A commonly used unit of volume is the liter (L). Smaller volumes can be measured in milliliters (mL). One milliliter is equal to 1/1000 of a liter. In the laboratory, the graduated cylinder is often used to measure the volume of liquids.

Temperature is measured with a thermometer. One unit of measurement for temperature is the degree Celsius (°C).

In this investigation, you will practice making measurements of the volume and temperature of a liquid.

Problem
How can you accurately measure the volume and temperature of a liquid?

Pre-Lab Discussion
Read the entire investigation. Then work with a partner to answer the following questions.

1. **Measuring** How many significant figures are there in the measurement shown in Figure 1?

   three significant figures

2. **Inferring** Why is it important to read the volume of water in a graduated cylinder by using the bottom of the meniscus?

   Because water curves upward at the edges of the cylinder wall, the bottom of the meniscus is nearer to being the true top of the water column. By using this same point each time a measurement is made, the measurement is less likely to be in error.

3. **Designing Experiments** Why should you leave the thermometer in beaker B when you add ice?

   Removing the thermometer from the water would expose the thermometer bulb to the air. The resulting change in the thermometer reading would be because of a temperature change other than that caused by the ice melting in the water.

4. **Measuring** If each mark on a thermometer represents 1°C, which part of a temperature measurement will be the estimated digit?

   tenths of a degree

Earth Science Lab Manual

Name _____ Class _____ Date _____

Materials *(per group)*

2 150-mL beakers
100-mL graduated cylinder   *Provide beakers with volume gradations.*
glass-marking pencil

2 Celsius thermometers
watch or clock   *Provide only non-mercury thermometers.*
ice cube

Safety 🥽 👕 ⚠️

Put on safety goggles and a lab apron. Be careful to avoid breakage when working with glassware. Note all safety alert symbols next to the steps in the Procedure and review the meaning of each symbol by referring to the Safety Symbols on page xiii.

Procedure

Part A: Measuring the Volume of a Liquid

⚠️ 1. Fill a beaker halfway with water.

2. Pour the water in the beaker into the graduated cylinder.

3. Measure the amount of water in the graduated cylinder. To accurately measure the volume, your eye must be at the same level as the bottom of the meniscus, as shown in Figure 1. The meniscus is the curved surface of the water.

4. Estimate the volume of water to the nearest 0.1 mL. Record this volume in Data Table 1.

5. Repeat Steps 1 through 4, but this time fill the beaker only one-fourth full of water.

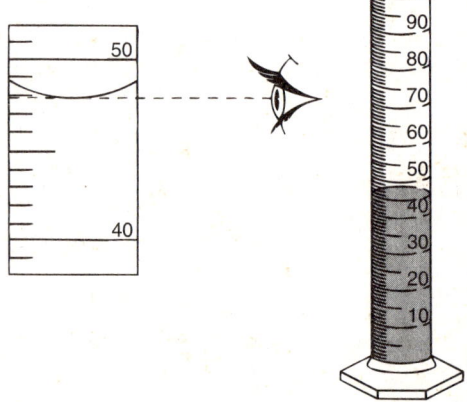

Figure 1

Part B: Measuring the Temperature of a Liquid

6. Use the glass-marking pencil to label the beakers *A* and *B*.

7. Use the graduated cylinder to put 50 mL of water in each beaker.

8. Place a thermometer in each beaker. In Data Table 2, record the temperature of the water in each beaker.   *Tell students not to stir the water with the thermometer.*

9. Carefully add one ice cube to the water in beaker B. Note and record the time.

10. After 1 minute, observe the temperature of the water in each beaker. Record these temperatures in Data Table 2.

11. After 5 minutes, observe the temperature of the water in each beaker. Record these temperatures in Data Table 2.

12. After the ice in beaker B has melted, use the graduated cylinder to find the volume of water in each beaker. Record these volumes in Data Table 3.

Name _____ Class _____ Date _____

Observations   Sample data are shown.

DATA TABLE 1

Measurement	Volume of Water (mL)
Half-filled beaker	47.0–53.0
One-fourth filled beaker	22.0–28.0

DATA TABLE 2

Beaker	Temperature at Beginning	Temperature After 1 Minute (°C)	Temperature After 5 Minutes (°C)
A	18.0–20.0	18.0–20.0	18.0–20.0
B	18.0–20.0	15.0–17.0	8.0–14.0

DATA TABLE 3

Beaker	Volume of Water at Beginning (mL)	Volume of Water at End (mL)
A	50.0	50.0
B	50.0	58.0–68.0

Earth Science Lab Manual

Name _____ Class _____ Date _____

Analysis and Conclusions

1. **Observing** What is the largest volume of a liquid that the graduated cylinder is able to measure? What is the smallest volume that the graduated cylinder is able to measure?

   100 mL; 1 mL

2. **Analyzing Data** Describe how the temperature of the water in beakers A and B changed during the investigation.

   The temperature of the water in beaker A did not change. The temperature of the water in beaker B decreased after the ice cube was added.

3. **Analyzing Data** How did the volume of water in beakers A and B change during the investigation? What do you think caused this change?

   The volume of water in beaker A did not change. The volume of water in beaker B increased because the melted ice added to the total amount of liquid water.

4. **Applying Concepts** Would you use a 100-mL graduated cylinder, a 25-mL graduated cylinder, or 10-mL graduated cylinder to measure 8 mL of a liquid? Explain your answer.

   To measure 8 mL of liquid, it is best to use a 10-mL graduated cylinder because it has the finest graduations and provides the most precise measurement.

Go Further

Some liquids do not form a meniscus in a graduated cylinder as water does. Use a 10-mL graduated cylinder to measure 8.0 mL each of water, isopropyl (rubbing) alcohol, and vegetable oil. Observe and draw the meniscus of each liquid. Label your drawings to show how you think the volume of each liquid should be measured. Explain why you think that the volumes should be measured in this way.

Vegetable oil and alcohol form a convex meniscus in a clean glass container. The volumes of these liquids are read at the top of the meniscus because most of the surface of these liquids is near the top of the meniscus (unlike water, whose surface is mostly near the bottom of the meniscus).

Name _____ Class _____ Date _____

Chapter 1 Introduction to Earth Science   Investigation 1A

The International System of Units (SI)

Introduction

Earth science, the study of Earth and its neighbors in space, involves investigations of natural objects that range in size from the very smallest parts of an atom to the largest galaxy. To measure and describe objects here on Earth as well as those far from our planet, Earth scientists use the **International System of Units (SI),** which is a decimal system of weights and measures. The base units of this system are shown in Data Table 1 on the next page. **Length,** which is the distance between two points, is measured in meters (m). The quantity of matter in an object, or **mass,** is measured in kilograms (kg). The amount of a substance is measured in **moles** (mol). The SI unit used to measure time is the second (s), and the unit for temperature is kelvins (K). Electric current is measured in units called amperes (A), and luminous intensity is measured in candelas (cd). Derived SI units, which are formed from combinations of the base units, are also shown in Data Table 1.

Prefixes are added to SI base units to indicate how many times more or what fraction of the base unit is present. For example, one thousand meters is a kilometer (km). One-thousandth of a meter is a millimeter (mm). Common metric prefixes and their symbols are shown in Data Table 2 on the next page.

To convert one SI unit into another, you simply move the decimal point either to the left or to the right. If you are changing a smaller unit into a larger unit, the decimal point is moved to the left. If you are converting a larger unit into a smaller unit, the decimal point is moved to the right. Figure 1 can be used to determine how many places the decimal point is moved during a conversion.

In this investigation, you will make measurements using SI units and convert SI units.

> Have students review the discussion of scientific inquiry in their textbooks. Explain that measuring in metric units is an important part of many inquiries in Earth science.
> **SKILLS FOCUS:** Measuring, Using Models, Comparing and Contrasting **TIME REQUIRED:** 45 minutes

Problem

What are common SI units, and how can they be converted and compared to other units of measure?

Pre-Lab Discussion

Read the entire investigation. Then work with a partner to answer the following questions.

1. **Measuring** Which SI unit would you use to measure the amount of juice in a glass?

   liter or milliliter

Earth Science Lab Manual • 11

Name _____ Class _____ Date _____

2. **Comparing and Contrasting** Which is larger—143.0 millimeters or 143.0 decimeters?

   143.0 dm

3. **Calculating** How many millimeters are 1.43 decimeters?

   143.0 mm

4. **Calculating** The average human body temperature in degrees Fahrenheit is 98.6. Use the Metric Conversion Table in the front of this manual to convert this value to degrees Celsius. Show your work.

   °C = (°F − 32°)/1.8; °C = (98.6°F − 32°)/1.8 = 37°C

DATA TABLE 1

SI Base Units			Derived Units		
Quantity	Unit	Symbol	Quantity	Unit	Symbol
Length	meter	m	Area	square meter	m^2
Mass	kilogram	kg	Volume	cubic meter	m^3
Temperature	kelvin	K	Density	kilograms per cubic meter	kg/m^3
Time	second	s	Pressure	pascal ($kg/m \cdot s^2$)	Pa
Amount of substance	mole	mol	Energy	joule ($kg \cdot m^2/s^2$)	J
Electric current	ampere	A	Frequency	hertz (1/s)	Hz
Luminous intensity	candela	cd	Electric charge	coulomb ($A \cdot s$)	C

DATA TABLE 2

Prefixes and Symbols		
Prefix[1]	Symbol[2]	Meaning
giga-	G	one billion times base unit (1,000,000,000 × base)
mega-	M	one million times base unit (1,000,000 × base)
kilo-	k	one thousand times base unit (1000 × base)
hecto-	h	one hundred times base unit (100 × base)
deka-	da	ten times base unit (10 × base)
deci-	d	one-tenth times base unit (0.1 × base)
centi-	c	one-hundredth times base unit (0.01 × base)
milli-	m	one-thousandth times base unit (0.001 × base)
micro-	μ	one-millionth times base unit (0.000001 × base)
nano-	n	one-billionth times base unit (0.000000001 × base)

[1] A prefix is added to the base unit to indicate how many times more, or what fraction of, the base unit is present. For example, a kilometer (km) means one thousand meters and a millimeter (mm) means one-thousandth of a meter.
[2] When writing in the SI system, periods are not used after the unit symbols and symbols are not made plural. For example, if the length of a stick is 50 centimeters, it would be written as "50 cm" (not "50 cm." or "50 cms").

Name _____ Class _____ Date _____

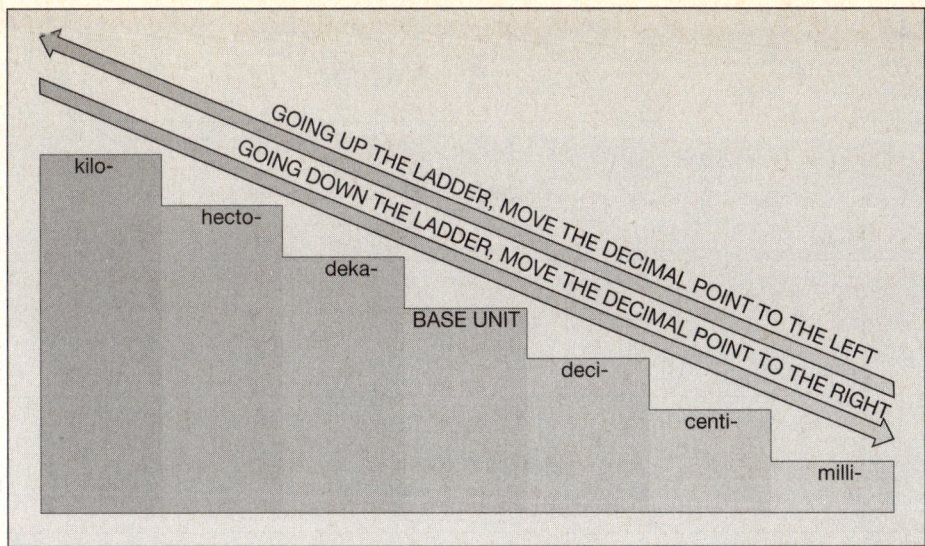

Figure 1 Metric Conversion Diagram

Materials *(per group of students)*
metric ruler
metric tape measure or meter stick
paper clip
nickel
paper cup
small rock
laboratory balance
large graduated cylinder
calculator
Metric Conversion Tables on page ix in this manual

You might want to set up various measuring stations that include both the appropriate instrument and instructions on what should be measured and allow students about 3–5 minutes per station. If necessary, review how to properly read a graduated cylinder and how to operate a balance. Make sure rock samples are small enough to fit into the graduated cylinder.

Safety
Be careful to avoid breakage when working with glassware.

Procedure
1. Work with a partner. Use the measuring tape or meter stick to measure either your height or your partner's height as accurately as possible to the nearest hundredth of a meter, or centimeter. Record your value in Data Table 3.
2. Change your value from Step 1 from centimeters to meters. Record this value in column 4 in Data Table 3. Use your calculator if necessary.
3. Repeat Steps 1 and 2 for all other measurements listed in Data Table 3 using the appropriate measuring device.
4. Use your data, the Metric Conversion Tables, and Figure 1 to answer the questions in the **Analysis and Conclusions** section.

Name _____ Class _____ Date _____

Observations

DATA TABLE 3

Height (cm)		Height (m)	
Length of page (cm)		Length of page (mm)	
Length of shoe (mm)		Length of shoe (m)	
Length of paper clip (mm)		Length of paper clip (km)	
Diameter of coin (mm)		Diameter of coin (μm)	
Volume of paper cup (mL)		Volume of paper cup (L)	
Volume of rock (mL)		Volume of rock (μL)	
Mass of paper clip (g)		Mass of paper clip (μg)	
Mass of paper cup (g)		Mass of paper cup (kg)	
Mass of coin (g)		Mass of coin (cg)	
Mass of rock (g)		Mass of rock (kg)	

Name _____ Class _____ Date _____

Analysis and Conclusions

1. **Calculating** Use Figure 1 to make the following conversions.
   a. 2.05 meters (m) = _____205.0_____ centimeters (cm)
   b. 1.50 meters (m) = _____1500.0_____ millimeters (mm)
   c. 9.81 liters (l) = _____98.1_____ deciliters (dL)
   d. 5.4 grams (g) = _____5400.0_____ milligrams (mg)
   e. 6.8 meters (m) = _____0.0068_____ kilometers (km)
   f. 4214.6 centimeters (cm) = _____42.146_____ meters (m)
   g. 321.50 grams (g) = _____0.3215_____ kilograms (kg)
   h. 70.73 hectoliters (hL) = _____707.3_____ dekaliters (daL)

2. **Calculating** Use the information on page ix in this manual to make the following conversions.
   a. On a cold day it was 8°F, or _____−13.3_____ °C.
   b. Ice melts at 0°C, which is _____32_____ °F.
   c. Room temperature is 72°F, or _____22.2_____ °C.
   d. On a hot summer day, the temperature was 35°C, which is _____95_____ °F.
   e. Water temperature in a warm shower is 27°C, or _____80.6_____ °F.
   f. Hot soup can be 72°C, which is _____161.6_____ °F.
   g. Water boils at 212°F, or _____373_____ K.

Work Space for Calculations

Earth Science Lab Manual ▪ 15

Name _____ Class _____ Date _____

3. **Analyzing Data** Use what you have learned about SI and the data you collected during this investigation to answer each of the following questions.

   a. The outdoor thermometer reads 28°C. Will you need your winter coat? _____no_____

   b. If your body temperature is 40°C, do you have a fever? _____yes_____

   c. The thermostat in your classroom reads 37°C. Are you shivering or perspiring? _____perspiring_____

   d. Can an average man weigh 90 kilograms? _____yes_____

   e. About how many meters tall is a fire hydrant? _____about 1 m_____

   f. Can one person drink 250 mL of coffee at breakfast? _____yes_____

   g. What is average room temperature in K? _____about 295 K_____

   h. About how thick is a dime? _____about 1 mm_____

   i. Can a typical bathtub hold 80 liters of water? _____yes_____

   j. Can a pork roast that weighs 18 grams feed a family of four? _____no_____

4. **Inferring** Explain why you think SI is used by most scientists around the world.

   SI is a standard system of weights and measure that is used in most fields of science

   because it is easy to use. The measurements made in SI units can be compared and evaluated.

   _____

   _____

   _____

Name _____ Class _____ Date _____

Chapter 1 Introduction to Earth Science   Investigation 1B

Using a Topographic Map to Create a Landform

Have students review the discussion of maps and mapping in their textbooks. **SKILLS FOCUS:** Measuring, Inferring, Using Models, Interpreting Diagrams/Photographs
TIME REQUIRED: 60 minutes

Introduction

One of the many tools used to study Earth's landscape is a **topographic map,** which represents Earth's three-dimensional surface in two dimensions. Topographic maps use **contour lines** to show **elevation,** or height above sea level, on a two-dimensional surface. A **contour line** joins points on a map that have the same elevation. Contour lines never intersect. The difference in elevation between one contour line and the next contour line is the **contour interval.**

In this investigation, you will interpret the contour lines on part of a topographic map and use them to create a three-dimensional model, or landform.

Problem

How can you use a topographic map to create a landform?

Pre-Lab Discussion

Read the entire investigation. Then work with a partner to answer the following questions.

1. **Applying Concepts** How does a topographic map show the elevation of the land?

   Topographic maps show differences in elevation through the use of contour lines.
   A contour line connects points on a map that have the same elevation.

2. **Forming Operational Definitions** In your own words, define the term *contour interval.*

   A contour interval is the difference in elevation between one contour line and the next contour line.

3. **Inferring** Why can contour lines never intersect?

   Any given point cannot be at two different elevations.

Earth Science Lab Manual ▪ 17

Name _____ Class _____ Date _____

4. **Interpreting Diagrams/Photographs** What kind of topography is indicated by contour lines that are very close together? By contour lines that are very far apart?

   Close contour lines indicate a steep topography, while widely spaced contour lines indicate a

   relatively flat area where elevation changes little with horizontal distance.

   _____

   _____

5. **Using Models** What is the advantage of creating a landform from a topographic map?

   A landform is a three-dimensional model that allows you to visualize actual changes in elevation.

   These changes are not shown on two-dimensional maps.

   _____

   _____

Materials *(per group)*
transparent shoebox with lid
nonpermanent, fine-lined marking pen
enlarged photocopy of part of a topographic map
cellophane or masking tape
modeling clay
metric ruler

For each group of students, make an enlarged photocopy of the Northeast, the Southeast, or the Midwest topographic map in Figure 3. The photocopies should be slightly smaller than the box lid, as shown in Figure 2.

Procedure
1. Place the topographic map provided by your teacher inside the lid of the plastic box so that the map can be seen through the top of the lid. Secure the map to the lid by using small pieces of tape near the map's corners.
2. Place the lid on your desktop or table. Use the nonpermanent marking pen to trace the topographic map onto the box lid. When you have finished tracing every contour line, remove the map from the inside of the lid and put the lid aside.

Name _____ Class _____ Date _____

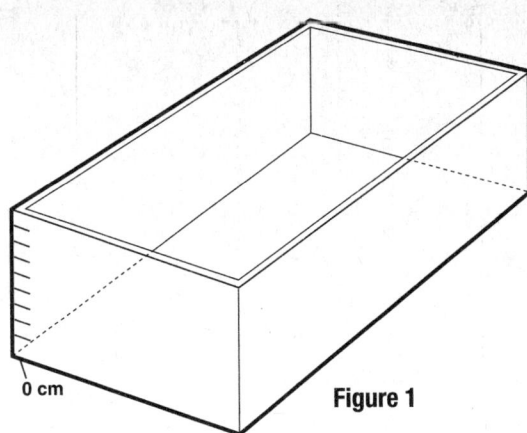

Figure 1

3. Use the marking pen and the metric ruler to make a centimeter scale along one of the vertical sides of the box, as shown in Figure 1.

4. Find the lowest elevation on the topographic map provided by your teacher. Write this elevation next to the bottom edge of the box.

5. Determine the contour interval of your topographic map. Each centimeter mark on the side of the box will represent the same vertical distance as the contour interval. Next to each centimeter mark, write the actual elevation in meters.

6. Use the modeling clay to make the first layer of the landform on your topographic map.

7. When you have finished the first layer of the landform, check it for accuracy. Do this by placing the lid on top of the box. Looking down through the lid, compare the landform with the corresponding contour lines on the map, as shown in Figure 2. Remove, add, or reshape the modeling clay, if necessary.

8. Repeat Steps 6 and 7 for each layer of the landform until all of the contour lines of the map have a corresponding layer on the landform.

9. After you have finished your model, use the following questions to record your observations.

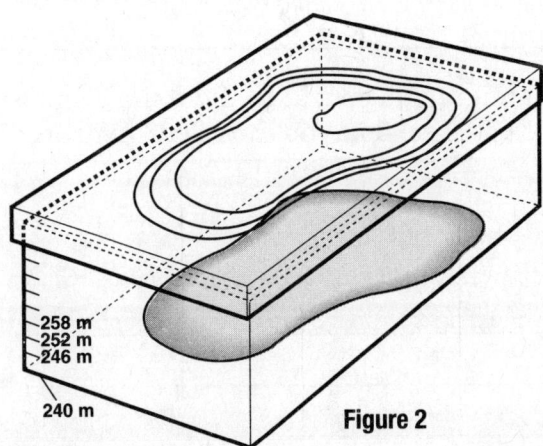

Figure 2

Earth Science Lab Manual ▪ 19

Name _____ Class _____ Date _____

Observations

1. What is the contour interval of the topographic map you used to make the landform?

   Answers will depend on the map used. The contour interval for the Southeast map is 3 m, that of the Northeast map is 30 m, and that of the Midwest map is 6 m.

2. Describe the shape of the landform you constructed.

   Answers will vary depending on the accuracy of students' models and the map used. The landforms should resemble one or two small hills.

3. How many meters above sea level is the base of your landform? How many meters above sea level is the top of your landform?

   Answers will vary depending on the map used. The base for the Southeast map is at 24 m, that of the Northeast map is 730 m, and that of the Midwest map is 240 m. The top of the Florida landform is greater than 33 m but less than 36 m; the top of the Vermont landform is greater than 910 m but less than 940 m; and the top of the Wisconsin landform is greater than 258 m but less than 264 m.

Analysis and Conclusions

1. **Using Models** What does your landform indicate about the region modeled?

   Answers will vary depending on the map used. The Southeast and Midwest maps show a fairly flat topography with gradual changes in elevation. These regions are plains. The map of the Northeast shows a much steeper landform, indicating that the region is hilly.

2. **Drawing Conclusions** What might you conclude about the overall Earth processes that shaped the region that you modeled? Explain your answer.

   Answers will vary depending on the map used. The Southeast and Midwest maps indicate gently sloping land, which is typical of surface processes such as weathering and erosion. The Northeast map indicates a hilly region, which could have been formed as the result of tectonic processes or even volcanism throughout the region.

Name _____ Class _____ Date _____

3. **Interpreting Diagrams/Photographs** Look at the topographic maps in Figure 3 on the next page. How does the topography of the Southwest map differ from the topographic features depicted in the other maps?

   The Southwest map shows a feature that is higher at the edges and drops in elevation toward the
   center. This is a map of a canyon or gorge. The other maps indicate hills or gently sloping plains.

4. **Comparing and Contrasting** Compare and contrast the topographic maps of the Southeast and the Southwest shown in Figure 3. Compare the processes that formed each landscape.

   Students should be able to infer that both landscapes formed as the result of moving water. The
   Florida landscape formed as water deposited, or dropped, bits of rocks and minerals as it moved
   over the land. The Grand Canyon, on the other hand, is the result of millions of years of erosion by
   the Colorado River.

5. **Inferring** Why do you think the landscape varies so much throughout the United States?

   The processes of erosion and deposition, as well as the interactions between tectonic plates, have
   had different effects on the land in various parts of the United States. These processes have resulted
   in the great geographical variety observed in different parts of the country.

Go Further

Repeat the investigation using one of the two other maps shown in Figure 3 on the next page. Or obtain a topographic map of a small region in your state and make a three-dimensional landform of the region. Be sure to choose a region that does not cover too large an area and does not have too great a change in elevation so that your landform will still be reasonably accurate. Before you attempt this activity, show your map to your teacher. When your teacher approves your map, carry out the investigation. When you have completed the model, record your observations.

Be sure that students choose a small enough area so that they are able to construct an accurate model. Have them avoid regions with extreme and irregular changes in elevation. Be sure that the map is enlarged sufficiently so that the contour intervals can be easily seen and modeled. Note that topographic maps of your state can be obtained from the United States Geological Survey (USGS) or your state survey.

Name _____ Class _____ Date _____

Southwest—Grand Canyon, Arizona
Contour interval: 50 meters

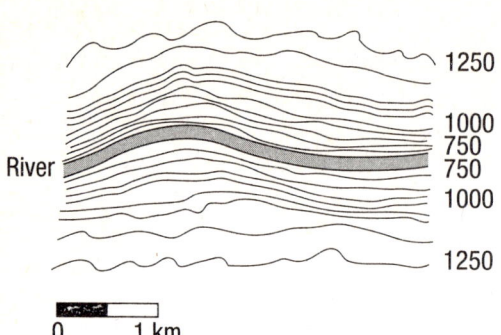

Northeast—Vermont
Contour interval: 30 meters

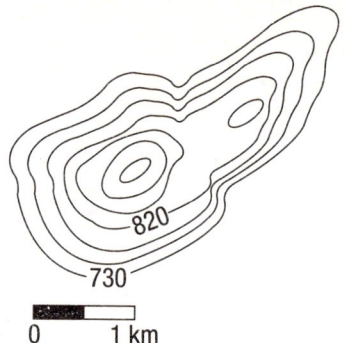

Southeast—Florida
Contour interval: 3 meters

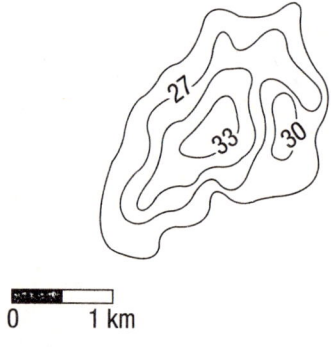

Midwest—Wisconsin
Contour interval: 6 meters

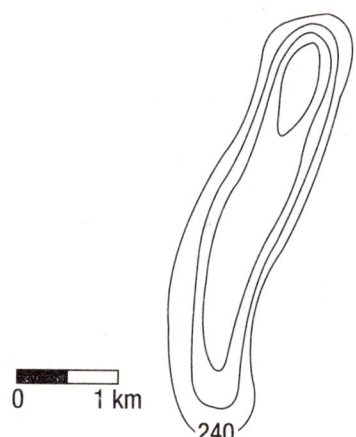

Figure 3 Some Hypothetical Topographic Maps from Different Parts of the United States

Name _____ Class _____ Date _____

Chapter 2 Minerals Investigation 2

Crystal Systems

Refer students to the discussion of crystal form in their textbooks. **SKILLS FOCUS:** Measuring, Using Models, Comparing and Contrasting **TIME REQUIRED:** 45 minutes

Introduction

Each of the more than 3800 minerals found on Earth has a distinct **crystal form** that is the result of the mineral's internal arrangement of atoms. These unique forms allow every mineral to be classified into one of six crystal systems, which are shown in the Data Table. Crystals in each system have a specific number of imaginary axes, which are indicated by the letters a, b, and c. These axes can vary in length and intersect at specific angles, designated by the Greek letters α, β, and γ. The angles between adjacent faces of a crystal can be measured with a simple instrument called a **contact goniometer.**

In this investigation, you will make a contact goniometer and use it to measure angles between adjacent faces of model crystals. You will then use your measurements to identify the system to which each model belongs.

Problem

How do crystal forms vary among the six different crystal systems?

Pre-Lab Discussion

Read the entire investigation. Then work with a partner to answer the following questions.

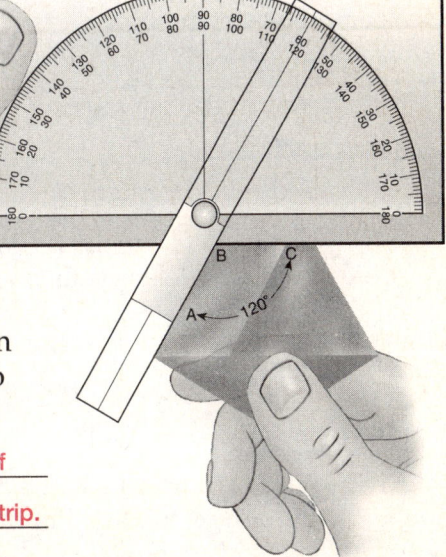

Figure 1

1. **Interpreting Diagrams** Look at Figure 1. In your own words, describe how you will use the goniometer to measure the angles between adjacent crystal faces.

   One face of the crystal being measured is flush with the edge of the cardboard while the adjacent face is flush with the plastic strip. The angle made by the two faces is indicated by the central line on the plastic strip and is read from the goniometer.

2. **Comparing and Contrasting** Look at the Data Table. How are minerals in the orthorhombic and monoclinic systems the same? How do they differ?

   All crystals in the orthorhombic system have three axes. Each axis is a different length. The axes intersect at 90° angles. All crystals in the monoclinic system have three axes that are all different lengths. Two of the axes in these crystals intersect at a 90° angle. The third axis intersects the other two axes at an angle greater than 90°.

Earth Science Lab Manual ▪ 23

Name _____ Class _____ Date _____

3. **Comparing and Contrasting** Look at the Data Table. Compare and contrast characteristics of minerals in the cubic and triclinic systems.

   All crystals in the cubic system have three axes that are all the same length. The axes intersect at 90° angles. All crystals in the triclinic system have three unequal axes that make oblique angles with each other.

4. **Interpreting Diagrams** Study the diagrams in the Data Table. Which system of crystals has three identical planes of symmetry? Explain your choice.

   Cubic crystals have three identical planes of symmetry. All three axes in these crystals are the same length, and all axes intersect at 90° angles.

Materials (per group)

small, thin protractor (like the one shown in Figure 1)
stiff, white cardboard (~ 20 cm × 10 cm)
clear, quick-drying glue
metric ruler
thin, clear, rigid plastic strip (~ 15 cm × 2 cm)
hole punch
2 gummed reinforcements
metal brad
copies of the model crystals on pages 26–28
scissors
cellophane tape

Make slightly enlarged photocopies of the model crystals on pages 26–28. Make enough copies so that each group of students has one complete set of models.

The optimal grouping for this activity is pairs. If time is limited, however, form groups of six and have each student in a group work with only one of the six models.

Before giving the strips to students, pre-cut and mark the central line on the plastic strips as indicated in Figure 1. Also pre-cut the pieces of cardboard using a paper cutter to ensure that the top and bottom edges of each piece are parallel.

Clear, cheap cellophane tape is best for this activity. Do not use removable tape.

Safety

Always use care when using scissors. Note the sharp object safety alert symbol next to Step 5 in the Procedure.

Procedure
Part A: Constructing the Contact Goniometer

1. Use a thin layer of glue to bind the protractor to the cardboard. Make sure the base of the protractor is about 1 cm from and parallel to the bottom edge of the cardboard, as shown in Figure 1. Wait for the glue to dry completely before trying to use the instrument.

2. Use the hole punch to make a hole in the center of the plastic strip, as shown in Figure 1.

3. Use the hole punch to make a hole in the cardboard along the bottom edge, as shown in Figure 1. Moisten and place the gummed reinforcements over the hole on either side of the cardboard.

4. Use the metal brad to attach the plastic strip to the cardboard. Adjust the brad so that the plastic strip moves freely.

Earth Science Lab Manual ■ 24

Name _____ Class _____ Date _____

Part B: Making and Measuring the Model Crystals

5. Carefully cut out each of the photocopied models provided by your teacher. Cut only along the outermost lines of each figure.

6. Using the Data Table as a guide, fold and tape the models to form six, three-dimensional crystals. After you make each fold, gently run your fingernail along the fold so that the angles between the faces in the model crystals will be more exact.

7. Use Figure 1 as a guide to measure the angles between adjacent faces in crystal A. Use your measurements to identify to which of the systems crystal A belongs. Record your choice on the appropriate line in the Data Table.

8. Repeat Step 7 for the other five model crystals.

Observations

DATA TABLE The Six Crystal Systems

System Name	Description	Typical Forms	
Cubic Model __A__	All crystals in the cubic system have three axes that are all the same length. The axes intersect at 90° angles.	$a = b = c$; $\alpha = \beta = \gamma = 90°$	
Tetragonal Model __C__	All crystals in the tetragonal system have three axes that intersect at 90° angles. Two of the axes are the same length.	$a = b \neq c$; $\alpha = \beta = \gamma = 90°$	
Hexagonal Model __F__	All crystals in the hexagonal system have four axes. Three of the axes are the same length and intersect at 60° angles. The fourth axis is longer or shorter than the other three.	$a = b \neq c$; $\alpha = \beta = 90°$; $\gamma = 120°$	
Orthorhombic Model __D__	All crystals in the orthorhombic system have three axes. Each axis is a different length. The axes intersect each other at 90° angles.	$a \neq b \neq c$; $\alpha = \beta = \gamma = 90°$	
Monoclinic Model __B__	All crystals in the monoclinic system have three axes each of which is a different length. Two of the axes in these crystals intersect at a 90° angle. The third axis intersects the other two at an angle greater than 90°.	$a \neq b \neq c$; $\alpha = \beta = 90° \neq \gamma$	
Triclinic Model __E__	All crystals in the triclinic system have three unequal axes that make oblique angles with each other.	$a \neq b \neq c$; $\alpha \neq \beta \neq \gamma \neq 90°$	

Name _____ Class _____ Date _____

Model Crystals

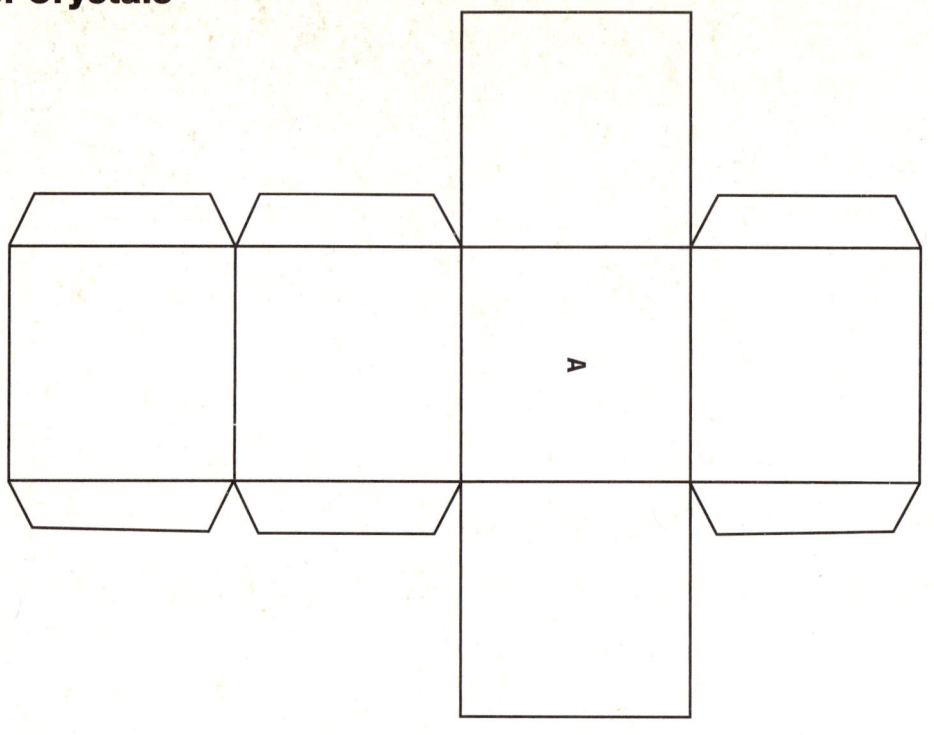

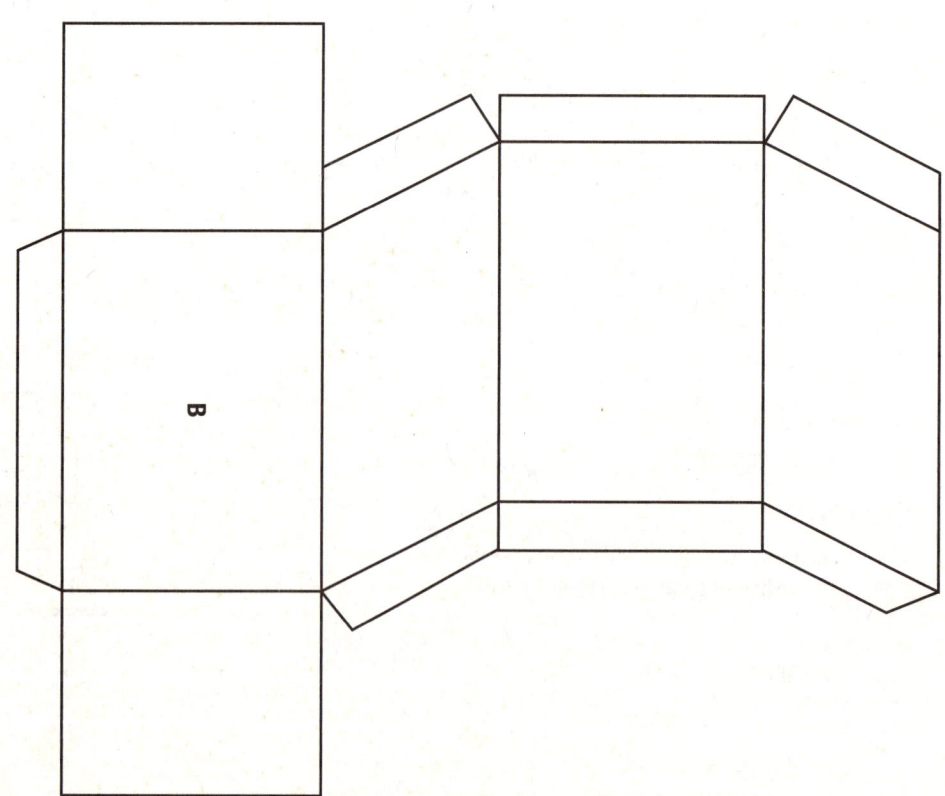

Name _____ Class _____ Date _____

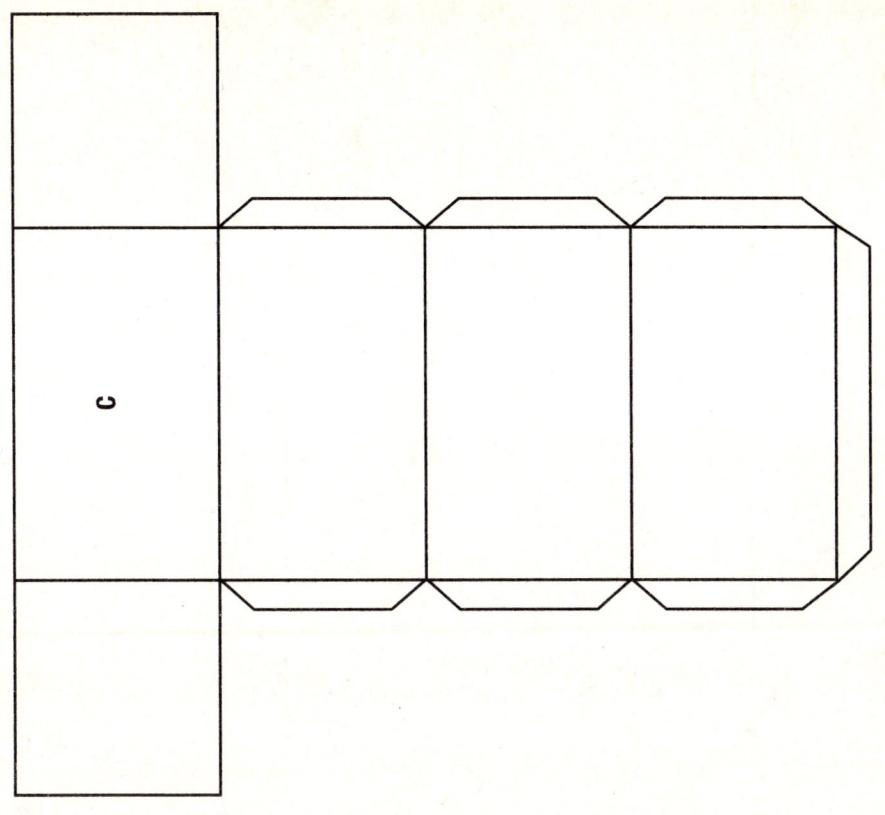

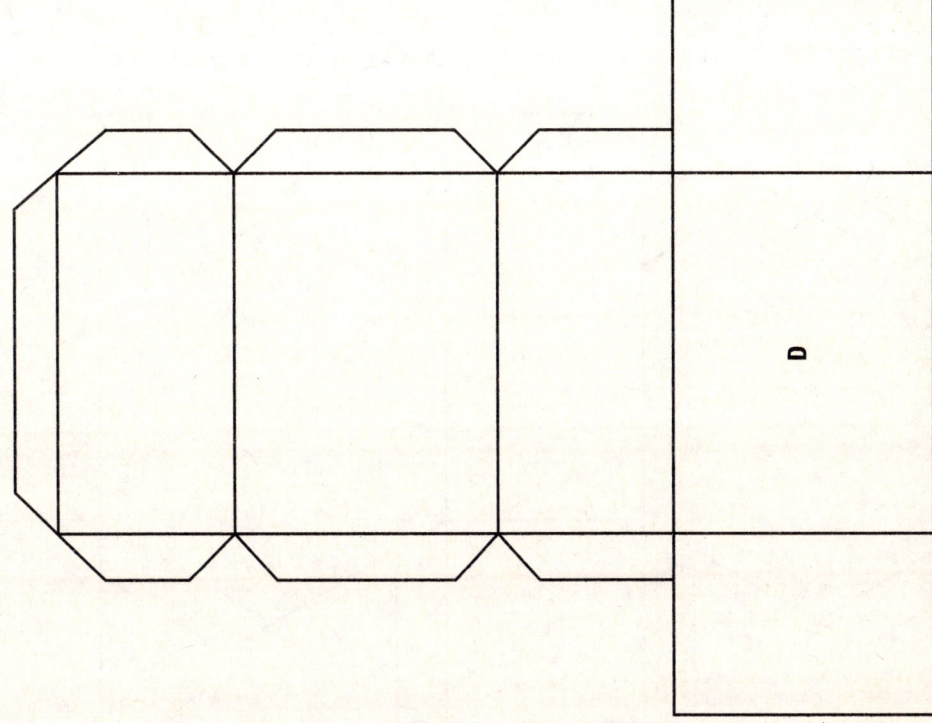

Earth Science Lab Manual ▪ **27**

Name _____ Class _____ Date _____

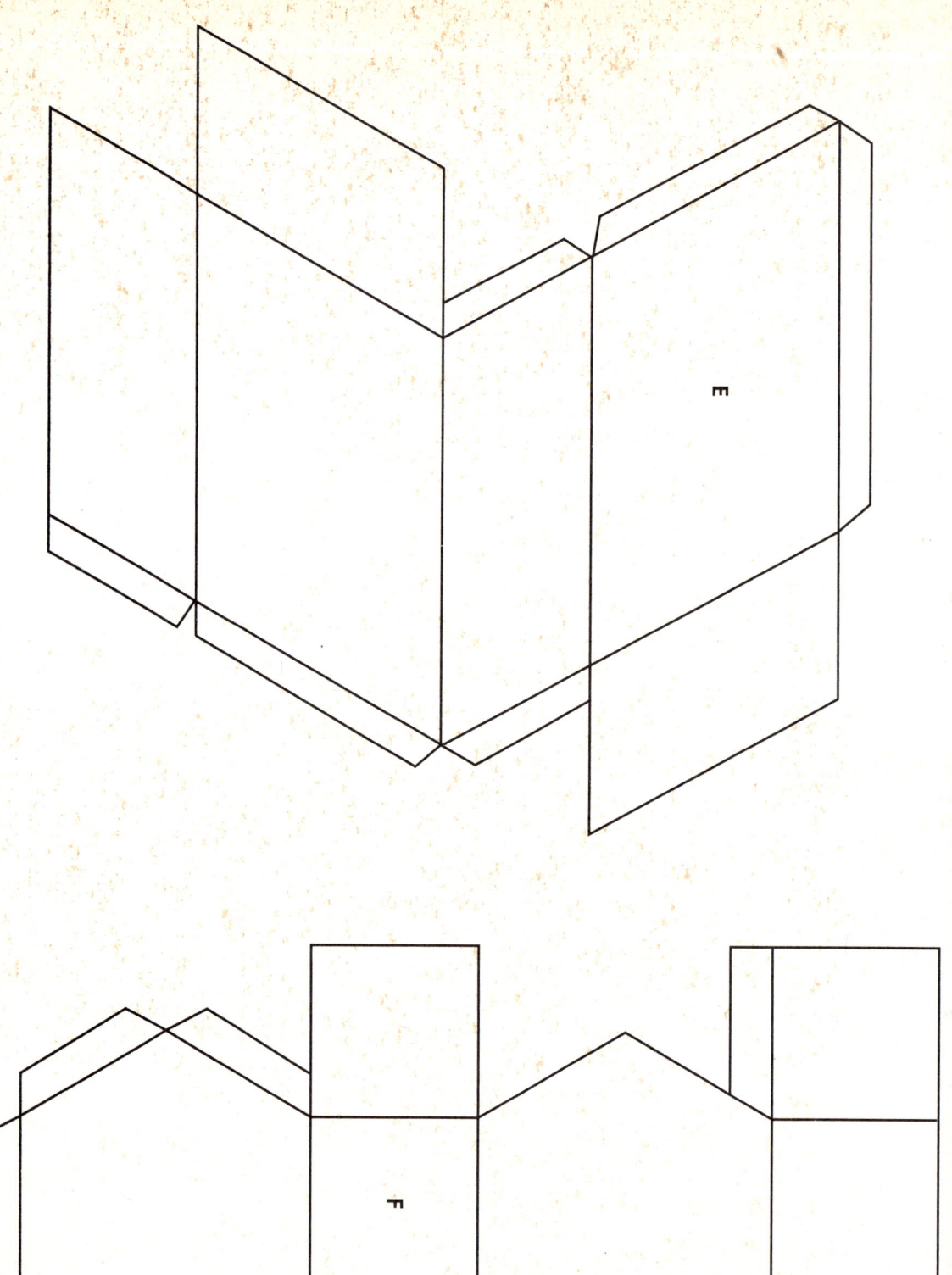

Earth Science Lab Manual ▪ 28

Name _____ Class _____ Date _____

Analysis and Conclusions

1. **Classifying** To which crystal system does each of the models belong?

   Model A is a cubic crystal; model B belongs to the monoclinic system; model C is a tetragonal crystal; model D belongs to the orthorhombic system; model E is a triclinic crystal; and model F belongs to the hexagonal crystal system.

2. **Comparing and Contrasting** How do the angles between major crystal faces differ among the various systems?

   In the cubic, orthorhombic, and tetragonal systems, the angles between adjacent faces are 90°. The angle between two adjacent faces in the hexagonal crystal is 60°. Angles between adjacent faces in the monoclinic crystal system are either about 116° or 64°. Angles between adjacent faces in crystals in the triclinic system are either about 102° or 88°.

3. **Drawing Conclusions** Why is crystal form probably the most useful property to use when identifying unknown minerals?

   Every mineral on Earth belongs to only one crystal system. Thus, crystal form is a very definitive property that can be used to identify an unknown mineral.

Go Further

Study the crystals shown in the photographs on Resource 2 in the DataBank. Use what you've learned in this investigation to identify the crystal system to which each mineral belongs.

Crystal A (fluorite): cubic
Crystal B (barite): orthorhombic
Crystal C (gypsum): monoclinic
Crystal D (quartz): hexagonal
Crystal E (rhodonite): triclinic
Crystal F (halite): cubic
Crystal G (feldspar): triclinic
Crystal H (wulfenite): tetragonal
Crystal I (sulfur): orthorhombic
Crystal J (beryl): hexagonal

Name _____ Class _____ Date _____

Chapter 3 Rocks

Have students review the information about rocks in their textbooks. **SKILLS:** *Observing, Classifying, Comparing and Contrasting* **TIME REQUIRED:** *45 minutes*

Investigation 3

Classifying Rocks Using a Key

Introduction

Note that it is better to have students complete this investigation on classification prior to their carrying out the Rock Identification lab in the student edition.

Recall that a **rock** is a naturally occurring, solid mass of minerals or mineral-like matter. Geologists classify rocks into three major groups based on how the rocks form. **Igneous rocks** form when molten material—**lava** or **magma**—cools either on Earth's surface or underground. **Extrusive rocks** form when lava cools quickly at or near Earth's surface. Extrusive rocks have either a fine-grained texture or a glassy texture. **Intrusive rocks** form as magma cools slowly farther beneath Earth's surface. This slow rate of cooling allows mineral grains to grow large, and such a rock is said to have a coarse-grained texture.

   Sedimentary rocks form when pieces of rocks, minerals, or organic matter—all of which are called **sediment**—are compacted and cemented. **Clastic rocks** are sedimentary rocks that are made of fragments of weathered Earth materials. The fragments might be fairly large, such as pebbles; somewhat smaller, such as grains of sand; or very small, such as grains of clay. **Chemical rocks** are sedimentary rocks that form when minerals settle out of solution. **Biochemical rocks** are sedimentary rocks that form as the result of organic processes.

   Metamorphic rocks are rocks that form when existing rocks are subjected to changes in pressure or temperature. They can also form when they are subjected to chemical solutions. Metamorphic rocks may be **foliated,** which means that the components are arranged in parallel bands, or **nonfoliated,** which means that the rock's components are not arranged in bands.

   In this investigation, you will observe rock **texture,** which is the shape, size, and arrangement of a rock's components. You will use rock texture and other properties to classify rocks using a key.

Problem

How can you use a key to classify rocks?

Pre-Lab Discussion

Read the entire investigation. Then work with a partner to answer the following questions.

1. Inferring What is the purpose of this investigation?

   The purpose of this investigation is to observe rocks' shapes, sizes, and arrangements of

   components in order to classify the rocks using a key.

2. Forming Operational Definitions In your own words, describe what is meant by a rock's *texture*.

   Answers will vary, but students should include that texture is a rock's pattern or lack of pattern

   made by the materials that make up the rock.

Earth Science Lab Manual ▪ 31

Name _____ Class _____ Date _____

3. Observing What distinguishes the two main types of igneous rocks? Explain your answer.

Grain size, which is determined by the rate of cooling, distinguishes intrusive igneous rocks from extrusive igneous rocks. Intrusive rocks have larger grains than extrusive igneous rocks.

4. Drawing Conclusions Chalk is made of tiny fragments of marine organisms. To which group of rocks does chalk belong?

Chalk is a biochemical sedimentary rock.

5. Classifying Suppose you observe a rock with distinct bands. What type of rock might this be? Can the rock also belong to another group of rocks? Explain your answer.

The banded rock might be a foliated metamorphic rock. The rock also could be a sedimentary rock that formed when sediments were deposited as the transporting agent—wind or water—lost energy and dropped the sediments in layers.

Materials *(per group)*

igneous rocks
sedimentary rocks
metamorphic rocks
bottle of dilute (1*M*) hydrochloric acid (HCl) with dropper
hand lens
paper towels
red pen or pencil

Order a variety of rock specimens from a scientific supply house. Samples of igneous rocks should include basalt, diorite, granite, pumice, and obsidian. Samples of sedimentary rocks should include limestone, conglomerate, chalk, sandstone, and shale. Samples of metamorphic rocks should include quartzite, gneiss, marble, slate, and schist. Choose specimens so that most of them can be classified using the key in Data Table 1. You may have to help students identify some of the specimens.

Prior to distributing the samples, identify each specimen with a letter either by using masking tape and a marker or by painting a small, solid, white circle on the rock and marking the circle. Keep a master list that identifies each sample.

Prepare a 1*M* solution of hydrochloric acid (HCl) by slowly adding 82 mL of concentrated (12*M*) HCl to 500 mL of distilled water in a 1000-mL flask. Then fill the flask with distilled water to the 1000-mL mark. Transfer the solution to standard laboratory bottles with droppers. Label each bottle.

Safety

Put on safety goggles, a lab apron, and plastic disposable gloves. Take care when using chemicals such as hydrochloric acid as they may irritate the skin or stain skin or clothing. Never touch or taste a chemical unless instructed to do so. Wash your hands thoroughly after completing this investigation. Note all safety alert symbols next to the steps in the Procedure and review the meaning of each symbol by referring to the symbol guide on page xiii.

Name _____ Class _____ Date _____

Procedure

1. Put on your safety goggles, disposable gloves, and lab apron.
2. Choose one of the rock samples provided by your teacher. Observe its texture, color, crystal size, and composition with and without the hand lens.
3. Use the Key to Rock Classification (Data Table 1) to classify the sample. Begin by reading the first question. Answer *Yes* or *No* based on your observations.
4. After the words *Yes* and *No*, you will find directions to proceed to another question, or you will discover to which group of rocks your specimen belongs. If you find directions to proceed to another question, go to that question, answer it, and follow the directions.
5. Continue working through the key in this way until you come to a statement that allows you to classify your rock sample. To answer Question 8 in Data Table 1, put the rock on a paper towel and place a single drop of HCl on the rock. **CAUTION:** *Always wear safety goggles and disposable gloves when working with chemicals.*
6. In Data Table 2, record the route that you take through the key, using the numbers of the questions. For example, your route could be "1—4—5—extrusive igneous rock." In the last column of the table, write the name of the rock group to which each sample belongs.
7. When you have classified all of the samples, remove and dispose of your rubber gloves and thoroughly wash your hands with soap and water.
8. Compare your classifications with those provided by your teacher. If you made a mistake in classifying any of the samples, put the correct answer in red next to your answer.

DATA TABLE 1

Key to Rock Classification	
1. Does the rock contain visible connecting crystals?	*Yes:* Go to Question 2. *No:* Go to Question 4.
2. Are all of the crystals the same color and shape?	*Yes:* The rock is a nonfoliated metamorphic rock (possibly marble or quartzite). *No:* Go to Question 3.
3. Are all of the crystals in a mixed "salt-and-pepper" pattern?	*Yes:* The rock is an intrusive igneous rock (possibly granite or diorite). *No:* The rock is a foliated metamorphic rock (possibly schist or gneiss).
4. Does the rock contain many small holes or have a uniform dark color?	*Yes:* The rock is an extrusive igneous rock (possibly pumice or basalt). *No:* Go to Question 5.
5. Is the rock glassy (does it resemble broken glass)?	*Yes:* The rock is an extrusive igneous rock (obsidian). *No:* Go to Question 6.
6. Does the rock have flat, thin layers that can be broken apart?	*Yes:* The rock is a foliated metamorphic rock (slate). *No:* Go to Question 7.
7. Does the rock contain pebbles, sand, or smaller particles that are cemented together?	*Yes:* The rock is a clastic sedimentary rock (possibly conglomerate, sandstone, or shale). *No:* Go to Question 8.
8. Does the rock fizz when dilute HCl is dropped on it?	*Yes:* The rock is chemical or organic sedimentary rock (limestone or chalk). *No:* Ask your teacher for assistance.

Earth Science Lab Manual

Name _____ Class _____ Date _____

Observations

DATA TABLE 2 Data will depend on the samples used for this investigation.

Letter of Sample	Route Taken	Group to Which Rock Belongs

Analysis and Conclusions

1. **Using Tables and Graphs** How difficult was it to use the key to classify your rock samples? What problems did you encounter?

   Answers will depend partly on how well the samples represent the rock type. Students may encounter difficulty if the samples appear to exhibit properties of more than one rock type or if they have subtle properties.

2. **Evaluating and Revising** For each sample that you incorrectly classified, retrace the route you took through the key. Do you need to correct your route? If so, write the correct route. If your route was correct, explain why you may have incorrectly identified the sample.

   If the route was correct, the rock may have been incorrectly identified because the key did not provide enough information to identify that sample.

Name _____ Class _____ Date _____

3. Making Generalizations How useful was rock color in classifying the rock samples? Explain your answer.

As with minerals, color is usually not very helpful for rock identification and classification. More than one type of rock can have the same color. Also, a certain type of rock may have more than one color depending on the minerals or other matter that compose it.

4. Making Generalizations Describe the overall texture of each of the major groups of rocks—intrusive, extrusive, clastic, chemical, foliated, and nonfoliated.

Answers will vary slightly but should be similar to the following: Intrusive rocks have coarse grains with interlocking minerals; extrusive rocks have fine grains and also can be glassy; the grain size in clastic rocks varies, but many clastic rocks appear as though their components have been glued together; chemical rocks often have fine grains; foliated rocks have components that are arranged in parallel bands; and nonfoliated rocks do not have components that are arranged in bands.

5. Comparing and Contrasting Which two of the rock samples were the easiest to classify? What properties made them easy to classify?

Answers will vary, but students may state that rocks with distinguishing properties were the easiest to classify. Those properties may include the glassy texture of obsidian, the fizzing of limestone when HCl is dropped on it, the bands or layers of foliated metamorphic rocks, or the cemented pebbles of sedimentary rocks.

6. Comparing and Contrasting Which two rock samples were the hardest to classify? Explain your answer.

Answers will vary but might include the following pairs of rocks: gneiss and schist, marble and quartzite, granite and diorite, and slate and shale. Each of these pairs of rocks has very similar textures.

Earth Science Lab Manual

Name _____ Class _____ Date _____

7. **Designing Experiments** How could you change the procedure of this investigation to produce better results?

Answers will vary, but students might say that different types of tests (other than the hydrochloric acid test) could have been done on the rocks to better classify them, or students might say that the classification key could have asked more questions.

Go Further

Collect at least ten different rocks samples from the area around your home or school. Label each rock with the location where you found it. Use the Key to Rock Classification (Data Table 1) to classify each of the rocks you gathered. **NOTE:** Always obtain permission **before** collecting rocks on public or private property. Only collect rocks when you are with a responsible adult.

Pair students and assign collecting areas to obtain the greatest variety of samples. Before assigning this part of the investigation, you may want to obtain geologic maps from your state's geological survey. These maps show the types of rocks exposed at the surface. Remind students to use care when collecting rocks, especially at outcrops. Students should always wear safety goggles and take care when using a rock hammer and moving along outcrops.

Name _____ Class _____ Date _____

Chapter 4   Earth's Resources   Investigation 4A

Recovering Oil

Have students review the information about petroleum traps and oil recovery in their textbooks. **SKILLS:** Predicting, Inferring, Comparing and Contrasting, Using Models **TIME REQUIRED:** 45 minutes

Introduction

When marine organisms die, they begin to decay and then become buried. Over millions of years, the decaying organisms undergo various physical and chemical changes, finally forming petroleum, or crude oil. The rock in which the petroleum forms is called the **source rock.** Over time, petroleum, along with natural gas, often moves from the source rock and collects in permeable rocks called **reservoir rocks.** Impermeable rocks, or **cap rocks,** such as shale prevent the oil and gas from moving further.

Petroleum in subsurface traps is often recovered by drilling a well into the reservoir. In some cases, the petroleum automatically rises as a result of changes in pressure in the reservoir. In other cases, however, pressure must be applied by methods that involve pumping water into the reservoir. This method allows the oil to be brought to the surface because petroleum is less dense than water and will float on top of it.

In this investigation, you will use various methods to recover the maximum amount of oil from a model well.

Problem

How does water affect the recovery of oil from a well?

Pre-Lab Discussion

Read the entire investigation. Then work with a partner to answer the following questions.

1. **Predicting** What do you predict will happen when you add water to the bottle after the first attempt to remove the oil has been made?

   Predictions will vary, but most students should state that additional oil will be brought to the

   surface when water is added to the model well.

2. **Predicting** Do you expect more oil to be removed from the model well when hot water or cold water is added to the well? Explain your answer.

   The addition of hot water will increase the temperature of the oil and make it less viscous,

   which will allow more of the oil to be "pumped" from the well.

Earth Science Lab Manual ▪ 37

Name _____ Class _____ Date _____

3. **Applying Concepts** How does adding water to an oil well change the pressure within the petroleum reservoir?

   The volume of fluids within the reservoir increases when water is injected into the reservoir.

   The added fluid causes an increase in pressure in the reservoir because the amount of space

   in the reservoir is fixed.

4. **Inferring** Different substances can be used to bring oil to Earth's surface. Why do you think water is used most often to get oil to flow from a well?

   Oil is less dense than water, so the oil will float on the water. Therefore, the oil will be forced

   to the top of the reservoir when water is injected into the oil well.

5. **Controlling Variables** What is the independent variable in this investigation?

   The water temperature is the independent variable.

6. **Controlling Variables** What is the dependent variable in this investigation?

   The amount of oil removed from the model well depends on the temperature of the water,

   which means the oil is the dependent variable in this investigation.

Materials *(per group)*
plastic bottle with spray pump
plastic tubing to fit the pump nozzle
pea-sized pebbles
2 graduated cylinders
200 mL vegetable or mineral oil
4 250-mL beakers
cold tap water
hot tap water
paper towels
clock or watch with second hand
soap to clean glassware

Ask one student in each group to bring in a large, transparent, clean, empty dishwashing-soap bottle. Ask another student in each group to supply a clean spray pump. Such pumps are commonly used in many types of hand lotion and hand soap bottles. Note that the pumps must be thoroughly flushed with hot tap water to remove any residue. Remind students to avoid contact with eyes and to wash hands after cleaning pumps.

Collect as much of the oil as possible for reuse in other classes. Provide a clean plastic container into which students are to dump the oil collected from the wells.

Earth Science Lab Manual ▪ 38

Name _____ Class _____ Date _____

Safety

Wear your safety goggles and a lab apron during this investigation. Take care when working with the hot water to avoid burns. Be careful to avoid breakage when working with glassware. Wipe up any water or oil spills immediately. Wash your hands thoroughly after completing this investigation. Note all safety alert symbols next to the steps in the Procedure and review the meaning of each symbol by referring to the symbol guide on page xiii.

Procedure

1. Put on your safety goggles and a lab apron.

2. Put pebbles into the plastic bottle until the bottle is half full, as shown in Figure 1.

3. Use one of the graduated cylinders to measure 100 mL of vegetable oil or mineral oil. Pour the oil into the bottle. **CAUTION:** *Be careful to avoid breakage when working with glassware. Also be careful not to spill any oil onto your clothing. Immediately wipe up any spills.*

4. Place the spray pump onto the plastic bottle while gently working its tube down through the pebbles. Attach the plastic tubing to the nozzle. Place the other end of the tubing into a 250-mL beaker, as shown in Figure 1.

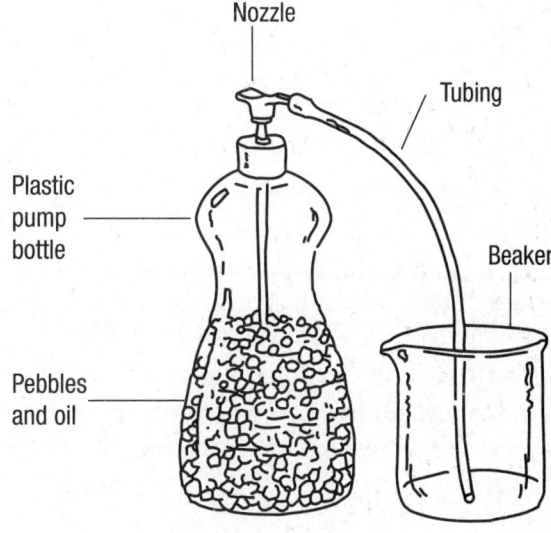

Figure 1

5. For 5 minutes, use the spray pump to remove as much oil as you can from the plastic bottle, which is your model well. Use the graduated cylinder to measure the amount of oil removed. Record this information in the Data Table.

6. Measure 80 mL of cold tap water in the second graduated cylinder. Add this water to the spray bottle. For 5 minutes, try to pump as much liquid as possible into a second beaker. Let the mixture stand for 1 or 2 minutes so that the oil and water separate. Carefully pour the oil layer that is on top of the water into the first graduated cylinder. In the Data Table, record the amount of oil you collected.

7. Rinse the bottle and pebbles with hot water to remove any excess oil. Dry them with paper towels. Place the used oil in the container designated by your teacher. Thoroughly clean the first graduated cylinder with soap and water. Add 100 mL of fresh oil. **CAUTION:** *Use extreme care when working with hot water to avoid burns.*

Name _____ Class _____ Date _____

8. Repeat Steps 2 through 5. After Step 5, add 80 mL of hot tap water to the spray bottle. For 5 minutes, try to pump as much liquid as possible into a second beaker. Let the mixture stand for 1 or 2 minutes so that the oil and water separate. Carefully pour the oil layer that is on top of the water into the first graduated cylinder. In the Data Table, record the amount of oil you collected.

9. Place the used oil in the container designated by your teacher. The water with traces of oil can be poured down the sink. **CAUTION:** *Use extreme care when working with hot water to avoid burns.*

10. Wash your hands thoroughly after completing this investigation.

Observations

DATA TABLE Sample data are shown below.

Oil Added to Model Well (mL)	Oil Removed Before Adding Water (mL)	Oil Removed After Adding Water (mL)
99.0–101.0 mL	50.0–65.0 mL	cold water 15.0–20.0 mL
99.0–101.0 mL	50.0–65.0 mL	hot water 25.0–30.0 mL

Analysis and Conclusions

1. **Using Models** How was your model reservoir like an actual petroleum reservoir? How was the model different?

   The model used petroleum (vegetable or mineral oil), the permeable reservoir rock (pebbles),

   and water. It did not have a source rock or cap rock.

2. **Analyzing Data** How did your results compare with your prediction?

   If students predicted that hot water would allow more oil to be removed from a well than cold water,

   their predictions were confirmed. If students predicted cold water, their predictions were

   not confirmed.

3. **Observing** Why were you unable to remove all the oil from the bottle when you pumped it in Step 5?

   Some of the oil adhered to the pebbles.

4. **Comparing and Contrasting** How did adding cold water to the model oil well affect the amount of oil removed from the well? Hot water?

   When either hot or cold water was added to the well, more oil could be pumped from the well.

   The hot water, however, allowed even more oil to be pumped from the well because the increase

   in temperature of the model reservoir made the oil less viscous and allowed more of it to be

   removed from the well.

Name _____ Class _____ Date _____

Chapter 4   Earth's Resources   Investigation 4B

Desalinization by Distillation

Have students review the information on the distribution of water on Earth in their textbooks.
SKILLS: Observing, Comparing and Contrasting, Using Models   **TIME REQUIRED:** Part A, 30 minutes setup; Part B, 20 minutes setup; 1–2 weeks for completion

Introduction

"Water, water everywhere, / Nor any drop to drink." This quotation is from *The Rime of the Ancient Mariner* by Samuel Taylor Coleridge, a poem that describes the fate of sailors in a boat that is stranded in the middle of the Pacific Ocean. The sailors are surrounded by water that they cannot drink. You might already know that both fresh water and seawater contain dissolved salts. But the amount of dissolved salts in a liter of seawater is much greater than the amount of dissolved salts in fresh water. Drinking seawater causes water to move out of cells in the body through a process called osmosis. This process can eventually cause dehydration and even death.

In regions where the supply of fresh water is limited, seawater can be treated to make it safe to drink. The general name for processes that remove salts from water is **desalinization.** One method for separating dissolved salts from water is called distillation. During the **distillation** process, a solution is heated until it vaporizes the liquid. The gaseous water is separated from the salt impurities. Then, as the vapor cools, the gas condenses to form liquid water that has been separated from the salts.

In Part A of this investigation, you will distill salt water by boiling it. In Part B of this investigation, you will use sunlight to vaporize the salt water.

Problem

How can you desalinize salt water?

Pre-Lab Discussion

Read the entire investigation. Then work with a partner to answer the following questions.

1. **Predicting** In Part A, how will the contents of the cooled flask differ from the contents of the heated flask?

   Predictions will vary, but most students should predict that the water in the heated flask contains
   dissolved salts. The water in the cooled flask will not contain dissolved salts.
   _____

2. **Inferring** In Part B, why is it important to keep the clear bottle away from direct sunlight?

   The temperature of the clear bottle must be kept low enough so that the water vapor will condense.
   _____
   _____
   _____

Earth Science Lab Manual ▪ 41

Name _____ Class _____ Date _____

3. **Formulating Hypotheses** In Part B, why is the bottle containing the salt water placed on a board? *Hint:* Assume that some water vapor condenses in the plastic tubing.

   If the bottles were at the same height, the tubing would be level and any condensed water vapor could flow back into the black-painted bottle, which would defeat the goal of the distillation.

4. **Inferring** Why are you able to separate the table salt from the water in the salt water?

   Salt water is a mixture that can be separated by physical means.

Materials (per group)

Have each group of students provide two clean, transparent, 2-L bottles from which the labels have been removed.

CAUTION: *Make sure all glassware used in this investigation is heat resistant.*

2 250-mL Erlenmeyer flasks
100-mL graduated cylinder
saltwater solution
aluminum foil (15 cm × 15 cm)
1 1-hole rubber stopper with glass tubing inserted
2 40-cm pieces of plastic tubing
2 pieces glass tubing, bent at right angles
500-mL beaker
crushed ice
matches
laboratory burner
ring stand and ring
wire gauze
heat-resistant gloves
2 2-L plastic bottles
duct tape
wooden board, at least 3 cm thick
glass-marking pen
2 100-mL graduated cylinders

CAUTION: *Inserting glass through a stopper can cause the glass to break. Wrap the glass tubing in a thick towel for your protection when gently inserting it in the stoppers. Use a lubricant such as glycerol. Attach the plastic tubing to the right-angle glass tubing in advance.*

Prepare the mixture by dissolving 50 g of salt for each liter of tap water.

Paint half of the plastic bottles black under a fume hood or outside.

Safety

Put on safety goggles and a lab apron. Be careful to avoid breakage when working with glassware. Never touch or taste any chemical unless instructed to do so. Use extreme care when working with heated equipment or materials to avoid burns. Be careful when using matches. Tie back loose hair and clothing when working with flames. Do not reach over an open flame. Note all safety alert symbols next to the steps in the Procedure and review the meaning of each symbol by referring to the symbol guide on page xiii.

Procedure

Part A: Distillation of Salt Water by Boiling

1. Put on your safety goggles and a lab apron.
2. Use the graduated cylinder to measure 100 mL of the saltwater solution and pour it into the flask that will be heated.

Name _____ Class _____ Date _____

3. Fill the beaker with crushed ice and position it away from the burner, as shown in Figure 1. Cover the second flask with aluminum foil, making sure that the glass tubing does not touch the bottom of the flask. Press the edges of the foil against the beaker. Place this flask in the crushed ice.

4. Connect the apparatus as shown in Figure 1.

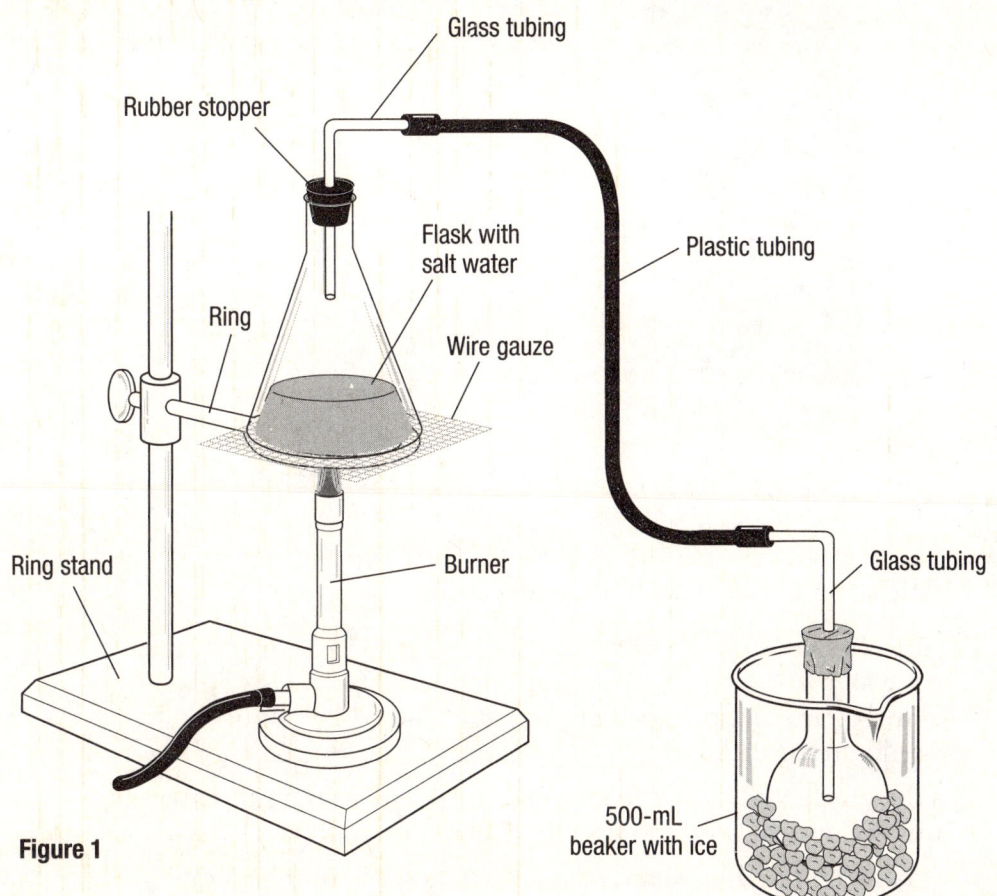

Figure 1

5. Using matches, carefully light the lab burner.

6. After about 8 to 10 minutes, carefully observe the flask in the beaker of crushed ice. **CAUTION:** *If necessary, carefully adjust the glass tubing so that it does not touch the liquid collecting in the flask.*

7. Continue collecting liquid in the cooled flask until about three-fourths of the liquid in the heated flask has been vaporized. Turn off the burner.

8. Observe the appearance of the contents of the cooled flask and the heated flask.

9. Put on heat-resistant gloves and remove the heated flask from the set-up. Allow the heated flask to cool. **CAUTION:** *Be careful handling equipment that has been heated. Hot glass looks like cold glass. Do not touch the heated flask or burner.*

10. Label and save the water in the cooled flask for the Go Further part of this investigation.

If the glass tubing is below the liquid, the condensed liquid may be drawn back up into the tubing. If this does occur, help students shake the liquid out of the tubing back into the cooled flask.

Earth Science Lab Manual ▪ 43

Name _____ Class _____ Date _____

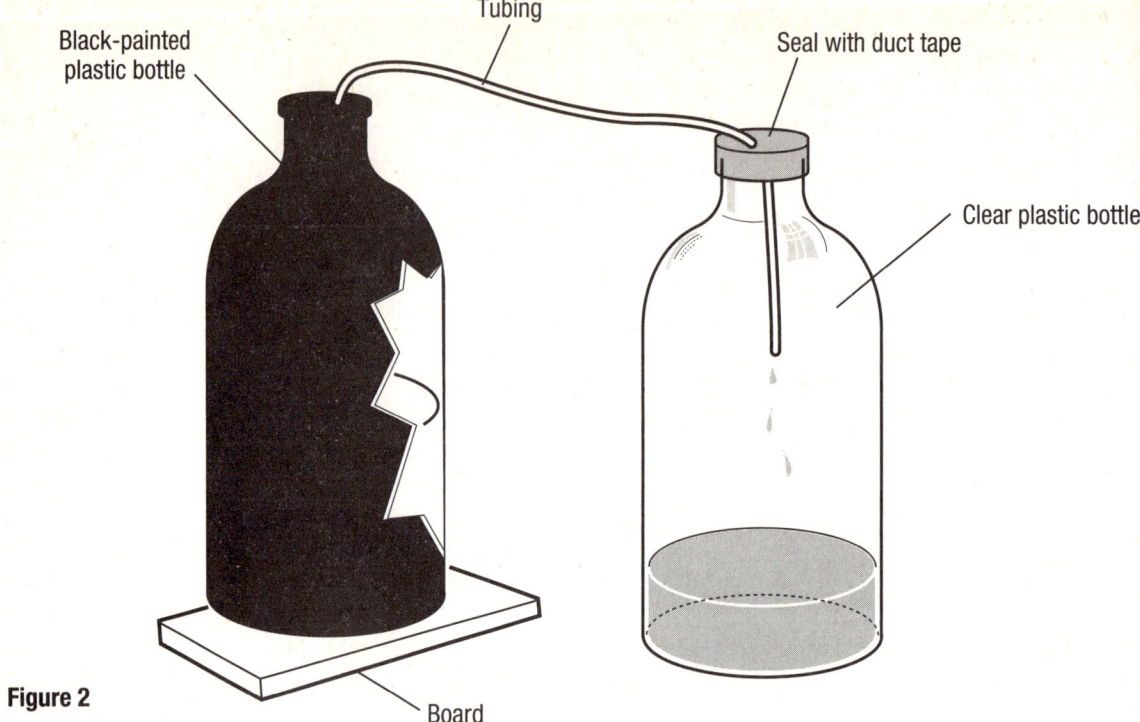

Figure 2

Part B: Distillation of Salt Water by Evaporation

11. Refer to Figure 2. Use the graduated cylinder to measure 100 mL of tap water. Pour the tap water into the clear bottle. Use a glass-marking pencil to mark the height of the water. Now pour the water down the drain.

12. Use the graduated cylinder to measure 100 mL of the saltwater solution. Pour the solution into the black bottle.

13. Insert one end of the plastic tubing into the neck of the black bottle and insert the other end into the neck of the clear bottle.

14. While you and another student hold the two bottles in place, ask a third student to wrap the duct tape around the necks of the bottles so that the plastic tubing is held tightly in place.

15. With another student, move the setup to a windowsill that has exposure to sunlight during at least half of the day. Set the black bottle on the wooden board so that it is at a higher position than the clear bottle. Arrange the bottles so that the black bottle receives as much sunlight as possible and the clear bottle is kept out of the sun as much as possible.

16. Check the bottles daily until at least 75 percent of the liquid has been transferred to the clear bottle.

17. Disconnect the apparatus and observe the appearance of both liquids.

18. Pour the liquid from the clear bottle into the cooled flask from Part A. You will use this water in the Go Further part of this investigation.

Earth Science Lab Manual ▪ 44

Name _____ Class _____ Date _____

Analysis and Conclusions

1. Comparing and Contrasting What is the major difference between the experiments in Part A and Part B? What effect does this difference have on the experiments?

The major difference between the experiments is the heat source—a gas burner in Part A

and sunlight in Part B. This difference affects the rate at which the distillation occurs.

2. Observing After you completed both parts of the investigation, how did the liquids in the heated flask and the black bottle differ from the liquids in the cooled flask and the clear bottle?

Students should state that the salt solution appeared cloudy as it became more concentrated or

that some salt crystals formed on the insides of the heated containers. If students did not carefully

observe the different liquids, however, they may state that there were no observable differences.

3. Applying Concepts Suggest at least one procedure you could use to demonstrate that the liquids in the heated flask and black bottle and the cooled flask and clear bottle have different compositions.

Students may suggest heating both liquids in open containers until all the water has evaporated.

There would be a solid residue from the salt solutions upon evaporation, but there would be no

residue with the distilled waters. Students may also suggest measuring and comparing the densities

of the liquids. The salt solutions would have a greater density that the distilled water. If students

suggest tasting the liquids, remind them that they should not taste any material in the laboratory.

4. Comparing and Contrasting List the advantages of each method used in this investigation for distilling water. What are some disadvantages of each method?

The advantages of distillation by boiling (Part A) are that the change occurs relatively quickly and

it is not dependent on the availability of sunlight. The advantages of using sunlight for distillation

(Part B) are that distillation requires less equipment, conserves nonrenewable energy sources, and

can be more easily accomplished in a non-laboratory environment. The disadvantages of distillation

by boiling are the expenses and the need for more equipment. The disadvantages of distillation by

sunlight are the dependence on the availability of sunlight and the amount of time it takes.

Earth Science Lab Manual

Name _____ Class _____ Date _____

5. **Applying Concepts** Based on your results, which of these methods do you think would be most cost effective should distillation be used to provide more of our freshwater needs?

   The second method (Part B) would be most cost effective because it uses solar energy,

   a renewable resource.

6. **Evaluating** Even though desalinization of seawater is used along some coastlines today to provide fresh water, there are some disadvantages to this process. List at least three disadvantages of desalinization.

   Answers will vary, but students might include the high costs; the potential problem of using or

   disposing of the salt residue; and the effects on ocean organisms, shoreline, and shallow water

   circulation that both removing the water and the desalinization plants might have.

Go Further

Without tasting the water, prove that the water in the cooled flask and the water in the clear bottle are fresh water. Pour 200 mL of salt water into a clean beaker. Pour 200 mL of the water you collected into another clean beaker. Predict what will happen if you place a fresh egg into each of the beakers. With your teacher's supervision, gently lower an egg into each beaker. Explain what you observe. Wash your hands thoroughly with soap or detergent after you complete the experiment.

An egg should float easily in salt water if the concentration is at least 5 mL of table salt in 200 mL of water. A fresh egg should not float in the distilled water because the density is less than that of salt water. Challenge students to predict how their observations of the two eggs will compare with placing an egg in tap water. Note that in some regions, tap water contains enough dissolved salts that a fresh egg will raise slightly in the water. Note, too, that the size of the eggs can affect the results.

Name _____ Class _____ Date _____

Chapter 5 Weathering, Soil, and Mass Movement   Investigation 5

Some Factors That Affect Soil Erosion

Introduction

Soil is a complex mixture of minerals, organic matter, air, and water. It is an integral part of Earth's lithosphere. Soil is home to billions of organisms, including many of the plants that people and other animals use as food. Soil stores nearly a quarter of Earth's fresh water, and it filters and decomposes many hazardous substances. Soil also plays an important role in Earth's cycles, including the nitrogen cycle and the carbon cycle.

You have learned that soil erosion depends on various factors including climate, slope, soil composition, and the type and amount of vegetation that grows in the soil. Some soil erosion is natural; however, much erosion is the result of human activities.

In this investigation, you will examine how slope affects soil erosion. You will also model some methods for reducing soil erosion.

Have students review the information on soil erosion and erosion control in their textbooks.
SKILLS: Observing, Measuring, Controlling Variables, Comparing and Contrasting
TIME REQUIRED: 45 minutes

Problem

How does slope affect soil erosion?

Pre-Lab Discussion

Read the entire investigation. Then work with a partner to answer the following questions.

1. **Posing Questions** Formulate a question that states the purpose of this investigation.

    Sample questions: Does an increasing slope angle increase the amount of soil eroded? Will less soil

    be eroded from a gentler slope than from a steeper slope?

    _____

    _____

2. **Controlling Variables** What are the independent and dependent variables in this investigation?

    The independent variable is the slope. The dependent variable is the amount of soil eroded.

    _____

    _____

    _____

Earth Science Lab Manual ▪ 47

Name _____ Class _____ Date _____

3. Inferring Why will you add some water to the sand before doing this investigation?

Adding water to the sand will prevent the sand from eroding too quickly and unevenly when water is poured over the soil.

4. Applying Concepts What are some methods that can be used to reduce soil erosion on slopes?

Answers will vary, but students might include terracing, building barriers to downhill soil movement, and planting vegetation.

Materials (per class)

large aluminum pans (4 to 5 depending on class size)
heavy-duty trash bag
scissors
duct tape
protractor
play sand (25-lb bag)
plastic drinking straws (3 to 4 per group)
ruler
wooden blocks of various heights (2 to 3 per group)
1-L plastic bottle
sprinkling can
bucket or other containers to collect waste water
clock or watch

Play sand can be obtained at most hardware or home and garden stores. Finer sand will work better than extremely coarse-grained sand. Students should pour sand into the pans and pack it down until there is uniform layer of sand about 5 cm thick.

Have students place a mark on the 1-L bottle at about 1000 mL in volume. Students should use this mark to deliver the same water volume with each trial.

Before attempting another trial using a different slope, students should remove as much of the water from the bottom end of the pan as possible and smooth out the sand in the pan. If time is not available for the student groups to complete several different slope angles, you may assign each group to investigate a different angle and have the groups share their data.

You can save time in the lab by cutting the trash bag into the 5-cm-wide strips. The strips should be cut in lengths that will extend across the pan before class.

The optimal size of each group doing this investigation is five to six students per group.

Safety

Put on safety goggles and a lab apron. Be careful to avoid breakage when working with glassware. Also be careful when handling sharp instruments. Wash your hands thoroughly with soap and water after completing this investigation. Note all safety alert symbols next to the steps in the Procedure and review the meaning of each symbol by referring to the symbol guide on page xiii.

Procedure

Part A: Modeling Erosion

1. Put on your safety goggles and a lab apron.
2. Pour sand into the aluminum pan and pack down the sand until there is a uniform layer of sand in the pan. The sand should be about 5 cm deep, as shown in Figure 1.

Earth Science Lab Manual • 48

Name _____ Class _____ Date _____

3. Lightly sprinkle a small amount of water over the sand to help pack down the sand.
4. Use one of the wooden blocks to prop up one end of the pan as shown.
5. Use the protractor to measure the angle the sand-filled pan has been raised by the wooden blocks. Record the angle for trial one in the Data Table.
6. Measure about 1000 mL of water into the sprinkling can. Slowly and evenly pour the water over the sand in the upper half of the pan.
7. Wait about 5 minutes for the erosion—movement of the sand—to stop. Observe how and where the sand erodes and how long it takes for the sand to begin eroding. Record your observations of the erosion.
8. Estimate the amount of sand (for example, one-third) that was eroded and deposited at the bottom end of the pan. Record the amount of sand eroded in the Data Table.
9. Scoop as much of the water out of the bottom end of the pan as possible and place the waste water in a bucket provided by your teacher. **CAUTION:** *Do not pour the sandy water down the drain.*
10. Repeat Steps 2 through 7 after placing more blocks under the pan or using a thicker block. Record your results for trial two in the Data Table.

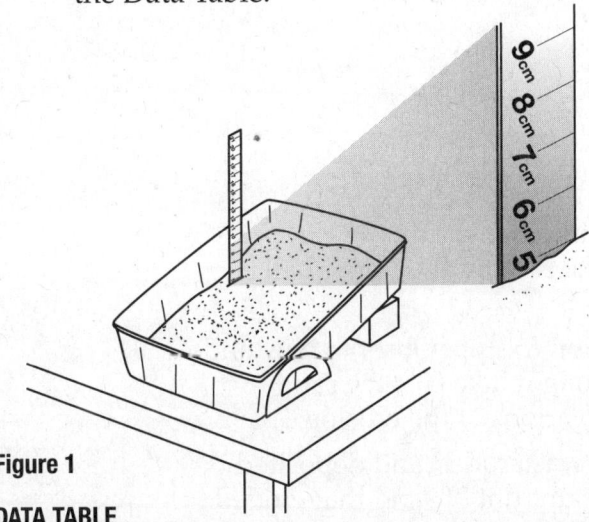

Figure 1

DATA TABLE

Trial	Angle of Pan	Amount of Sand Eroded	Observations
1			
2			
3			

Earth Science Lab Manual • 49

Part B: Modeling Erosion Barriers

11. Remove as much of the water from the pan as possible, using the bucket provided by your teacher. **CAUTION:** *Do not pour the sandy water down the drain.*

12. Smooth and pack down the sand. Measure the angle of the pan. Record the angle of the pan in the Data Table.

13. Use the scissors to cut the trash bag into a 5-cm-wide strip that is long enough to extend across the width of the aluminum pan.

14. Cut 3 to 4 plastic drinking straws in half or in lengths of about 10 cm.

15. Place 5 to 6 plastic straw sections in a line that extends across the middle of the sand-filled pan. Push the straw sections as far down into the sand as possible. Space the straw sections evenly across the pan, with one straw placed close to each side of the pan.

16. Tape the trash bag strip onto the top of the straw sections that stick up out of the sand, as shown in Figure 2. This strip represents an erosion barrier.

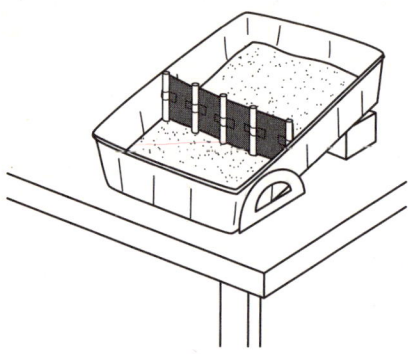

Figure 2

17. Measure about 1000 mL of water into the sprinkling can. Slowly and evenly pour the water over the sand in the upper half of the pan.

18. Wait about 5 minutes for the erosion to stop. Observe how and where the sand erodes and how long it takes for the sand to begin eroding. Record your observations of the erosion.

19. Estimate the amount of sand that was eroded and deposited in the bottom of the pan. Record the amount of sand that eroded for this trial.

20. Pour the sand and water into the bucket. **CAUTION:** *Do not pour sandy water down the drain.* Wash your hands thoroughly when you have finished the investigation.

Name _____ Class _____ Date _____

Analysis and Conclusions

1. **Comparing and Contrasting** In which trial of this investigation did the most soil erode?

   The greatest amount of soil was eroded when the pan was propped at the greatest angle.

2. **Relating Cause and Effect** Use your results to explain how slope affects the erosion of soil.

   All other factors being equal, the greater the slope, the more soil will be eroded.

3. **Inferring** In the repeated trials of the investigation, what variable or condition changed? How did this affect the amount of erosion?

   In repeated trials, the packed-down sand contained more water than at the start of the investigation.

   The wetter sand held together better and resulted in slightly less erosion than the drier sand.

4. **Drawing Conclusions** How did the erosion barrier affect the amount of erosion?

   The erosion barrier reduced the amount of erosion by acting as a baffle to catch the eroding

   sediment.

5. **Applying Concepts** Where could erosion barriers like the one you constructed in the investigation be used to reduce or prevent erosion?

   Answers will vary, but students might include construction sites, steep hillsides prone to erosion,

   and edges of newly plowed fields.

6. **Applying Concepts** How do you think soil erosion compares in dense forests, farms with crops planted in rows, and grasslands?

   The amount of soil eroded from densely vegetated areas—such as grasslands and forests—are

   much less than the soil that erodes from cultivated land.

Earth Science Lab Manual

Name _____ Class _____ Date _____

7. **Controlling Variables** How could you change the procedure in this investigation to obtain better results?

Answers will vary but might include using a soil mixture instead of sand and determining the mass of the sand/soil eroded by allowing the sediment to run out of the pan through an opening in the bottom end of the pan and then measuring the amount of sediment.

Go Further

Design a setup similar to the one used in the investigation that will explore the effect of different rates of rainfall. Pour water in the pan at faster and slower rates or in one location rather than over the entire pan and then compare the amount of erosion. Generate a detailed procedure and show it to your teacher. Once you have approval, predict what you think will happen and carry out the experiment. Dispose of the sand and water in a bucket provided by your teacher. Wash your hands thoroughly with soap or detergent after you complete the experiment.

Students will find that faster rates of rainfall will result in higher amounts of erosion. Pouring the water into the pan in one location will probably result in a channel developing in the sediment.

Name _____ Class _____ Date _____

Chapter 6   Running Water and Groundwater

Investigation 6A

Rivers Shape the Land

Have students review the information on streams and topographic maps in their textbooks. **SKILLS:** Observing, Measuring, Calculating, Interpreting Diagrams/Photographs
TIME REQUIRED: 45 minutes

Introduction

A **topographic map** is a model that represents Earth's three-dimensional surface in two dimensions. **Contour lines** on a topographic map connect points of equal elevation. The difference in elevation between adjacent contour lines is the **contour interval.** The vertical distance between the lowest and highest points shown on a topographic map is called **relief.** Closely spaced contour lines indicate a steep slope, while contour lines that are farther apart indicate a gentle slope. Like all maps, topographic maps have a **scale** that is used to show how horizontal distances on the map are related to actual distances shown on the map. Also like most other maps, topographic maps show a bird's-eye, or top, view of an area. Geologists often make **topographic profiles,** or side views, of an area to better visualize the change in elevation of an area.

The topographic map you will use in this investigation is of an area in Louisiana that is changed by the Red River and its tributaries. Like all streams, the Red River erodes materials from its channel and deposits these sediments elsewhere. Look at Resource 8 in the DataBank. Note that the width of the river's **floodplain,** or valley floor, in this part of Louisiana is shown by the solid line marked **A.** Other features shown on the map include levees, meanders, point bars, a yazoo tributary, an oxbow lake, and backswamps. A **levee** is a landform that is parallel to a stream and forms when the stream overflows its banks. Sediment that accumulates on the inside of a **meander,** or curve in the stream, is a **point bar.** A **yazoo tributary** is a tributary that flows parallel to the main river on a floodplain. An **oxbow lake** is a branch of a stream that becomes cut off from the main stream. **Backswamps** are poorly drained areas on a river's floodplain.

In this investigation, you will use a topographic map to answer questions about the Red River. You will also make a topographic profile of a section of the map.

Problem

How does a river change the land over which it flows, and what do these features look like on a topographic map?

Pre-Lab Discussion

Read the entire investigation. Then work with a partner to answer the following questions.

1. **Interpreting Diagrams/Photographs** What is the contour interval of this topographic map?

   20 feet
   _____

Name _____ Class _____ Date _____

2. **Interpreting Diagrams/Photographs** Approximately how many inches on the map represent 5 miles on the surface?

   The map scale is 1 inch equals 1 mile. So, 5 miles is about equal to 5 inches on the map.

3. **Interpreting Diagrams/Photographs** In which part of the map is the topography steeper? Explain your answer.

   The topography in the southern part of the map is steeper than the topography in the northern part of the map. The steepness is indicated by the closely spaced contour lines.

4. **Interpreting Diagrams/Photographs** Find the small, red number 10 in the south central part of the map. What is the total relief between the small stream and the small hill to the northwest?

   The highest contour line is at 200 feet; the lowest contour line represents 160 feet. The streambed is below 160 feet but above 140 feet. The relief is about 59 feet.

5. **Interpreting Diagrams/Photographs** Use the contour lines to determine the direction in which the small stream mentioned in Question 4 is flowing.

   The stream is flowing almost directly south.

6. **Interpreting Diagrams/Photographs** Recall that gradient is the slope or steepness of a stream channel. Describe the gradient of the Red River in this part of Louisiana.

   The gradient is very low because the area is very flat.

Materials (per group)
pencil
metric ruler
plain white paper, 1 sheet
Resource 1 in the DataBank
Resource 8 in the DataBank

Procedure
1. To make a topographic profile, place the sheet of white paper along the vertical line marked **P-P'** in the eastern part of the map.

You may want to provide students with copies of the topographic map, Resource 8, in the DataBank. On an overhead transparency, make a simple topographic map of a hill by drawing a series of closed contours. Label the contours using a contour interval of 5 or 10 feet. Draw a line through the center of the hill. Place the transparency on the overhead and use another blank transparency to demonstrate how to make a topographic profile. Refer to the Procedure that follows if you are unsure of how to construct the profile.

Earth Science Lab Manual ▪ 54

2. Mark each place where a contour line intersects the edge of the paper, as shown in Figure 1. Record the elevation of the contour line next to each mark on the paper. Note that in the southern portion of the map the contour lines are very closely spaced. Mark and label only every third or fifth line.

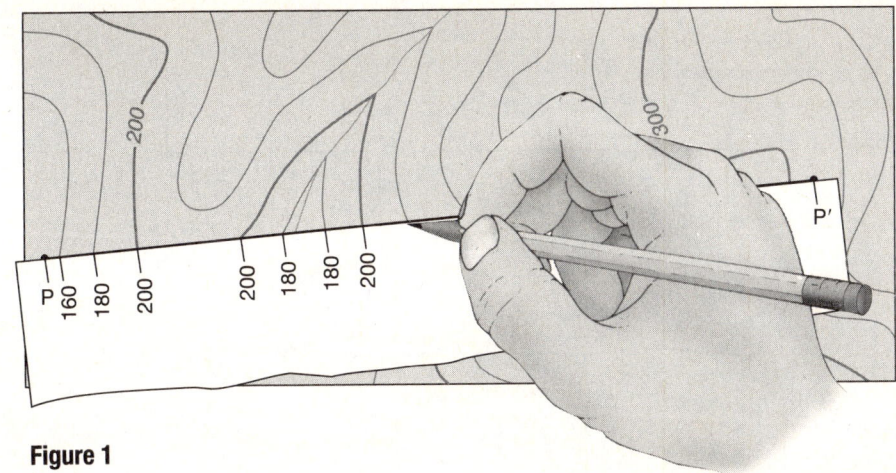

Figure 1

3. Based on the values you marked along the piece of paper, decide on a vertical scale for your profile. Include a value slightly lower than the lowest point recorded and a value slightly higher than the highest point recorded. Use the ruler to mark this scale on Figure 2.

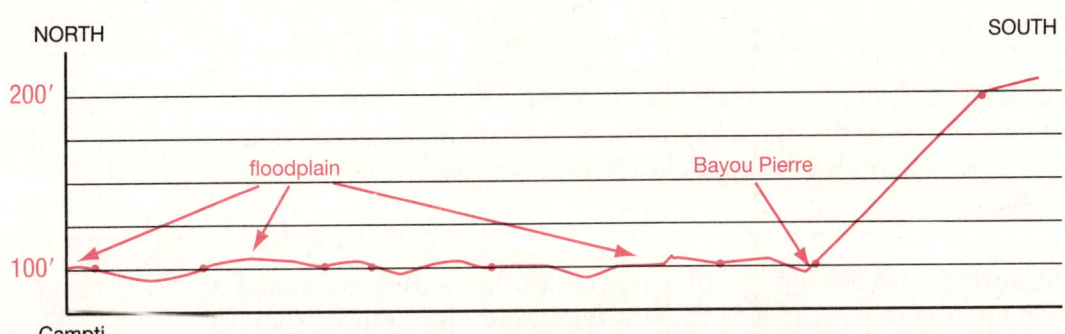

Figure 2

4. Lay your marked paper along the base of Figure 2. Wherever you have marked a contour line on the paper, place a dot directly above the mark at the appropriate elevation, as shown in Figure 3 on the next page. Note that the values shown in Figure 3 are not the same as the values you marked on your paper.

5. Connect your points with a line. Again, note that your profile will not be the same as the one shown in Figure 3.

6. Label the floodplain area and Bayou Pierre on your topographic profile.

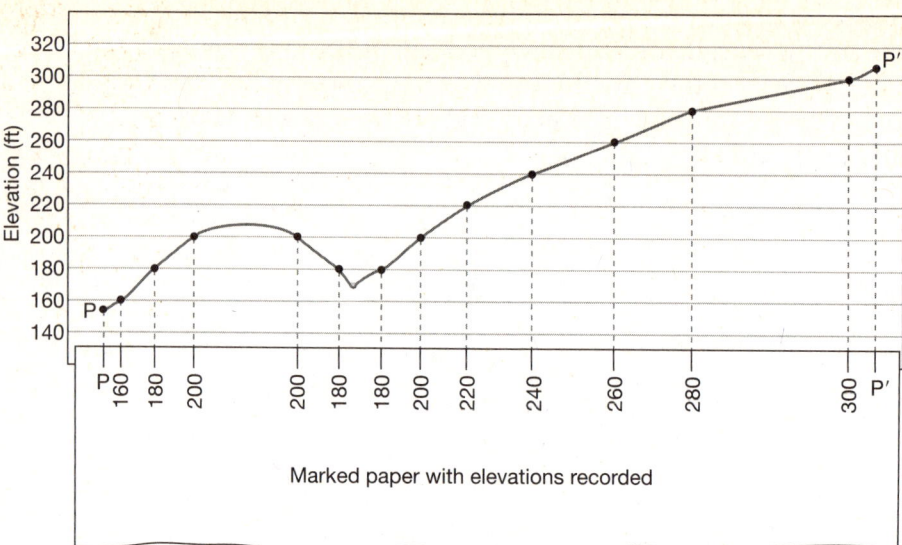

Figure 3

Analysis and Conclusions

1. **Interpreting Diagrams/Photographs** Describe your profile.

   Answers will depend on how accurately students transferred the information from the map to the profile. However, most profiles should show a flat region that extends over nearly 75 percent of the profile, and at Bayou Pierre, elevation increases dramatically. Profiles should resemble those shown in the teacher's edition of this manual on page 55.

2. **Calculating** Recall from your textbook that ultimate base level is sea level. Use your profile to determine approximately how many feet the Red River's floodplain is above ultimate base level.

   about 100 to 120 feet above sea level

3. **Measuring** Use the topographic map to estimate the percentage of the map area that is the Red River's floodplain. Remember that the line on the map labeled **A** marks the width of the floodplain.

   Approximately 75 percent of the map area is floodplain.

4. **Forming Operational Definitions** What do you think the area marked by the dashed lines labeled **B** is called? *Hint:* Look at the river's path to help you answer this question.

   The area between the dashed lines marked B is the river's meander belt.

Earth Science Lab Manual

5. **Comparing and Contrasting** Contrast the width of the area between the dashed lines labeled **B** with the width of the Red River's floodplain.

The width of the floodplain is several times wider than the width of the meander belt.

6. **Interpreting Diagrams/Photographs** Identify the structure on the map that is labeled **C**. How does such a structure form?

The structure labeled C is an oxbow lake that formed when one meander overtook another to form a ring of water on the river's floodplain. Sediments deposited around the ring isolated the ring, forming the lake.

7. **Interpreting Diagrams/Photographs** Identify the structure on the map that is labeled **D**. Explain how it formed.

The structure labeled D is a point bar that formed as sediment was deposited on the inside of the river's meander.

8. **Interpreting Diagrams/Photographs** What kind of feature is labeled **E** on the topographic map?

The structure labeled E is a yazoo tributary.

9. **Interpreting Diagrams/Photographs** What kind of feature is labeled **F** on the topographic map? How might this feature have formed?

The structure labeled F is a backswamp, or poorly drained area on a floodplain. A backswamp forms when levees prevent water from draining into the main river channel.

10. **Interpreting Diagrams/Photographs** What kind of feature is labeled **G** on the topographic map?

This curve in the river is called a meander.

Name _____ Class _____ Date _____

11. Inferring How could you change your technique in this investigation to obtain better results?

Answers will vary, but students might state that they could have marked every other contour line rather than every third or fifth.

Go Further

Look again at the map. Write several sentences to explain how the Red River is changing its channel in this part of Louisiana and how the floodplain might change over the next 100 years.

Students should be able to conclude that lateral erosion rather than downcutting is the dominant activity of the Red River in this part of Louisiana. Assuming that this erosion by the Red River continues without interruption, the floodplain will widen over time.

Name _____ Class _____ Date _____

Chapter 6 Running Water and Groundwater

Investigation 6B

Modeling Cavern Formation

Introduction

Caverns are chambers that form when acidic groundwater slowly erodes limestone formations beneath Earth's surface. These subsurface chambers, which are commonly called caves, form at or below the water table as groundwater flows along joints and bedding planes in the rocks. Many of the minerals dissolved by the groundwater are eventually deposited as dripstone features. **Stalactites** are icicle-like stone pendants that hang from the ceilings of most caverns. **Stalagmites** are features that form when a saturated water solution drops onto cavern floors and evaporates. This process leaves behind minerals that accumulate until they reach the cavern ceiling. Over time, a downward-growing stalactite and an upward-growing stalagmite might join to form a column.

In Part A of this investigation, you will model the formation of a cavern. In Part B, you will model how stalactites and stalagmites form.

Have students review the information on groundwater erosion and deposition in their textbooks. **SKILLS FOCUS:** Observing, Inferring, Comparing and Contrasting, Using Models **CLASS TIME:** 45 minutes to perform both parts of the investigation plus 3 to 5 minutes on 7 to 10 consecutive days to observe the model dripstone features

Problem

If you do not have sufficient space for the setups in Part B of this investigation, perform this part of the lab yourself in front of the class. Place the setup in an area easily accessible to students. Allow them to observe the setup daily and to answer related questions. Students should still work in small groups to perform Part A of the lab.

How can you model the formation of caverns and their deposits?

Pre-Lab Discussion

Read the entire investigation. Then work with a partner to answer the following questions.

1. **Posing Questions** Write one question that summarizes the purpose of both parts of this investigation.

   Sample answers: How does groundwater form and change caverns? What are the effects of groundwater erosion and deposition on subsurface limestone?

2. **Predicting** Based on the information given in the procedure for Part A, predict what you think will happen to the clay and its contents.

   Answers will vary, but most students should be able to predict that the sugar cubes or cough drops, which represent limestone, will at least partially dissolve to form voids in the ball of clay. The clay, which represents insoluble rocks, will not dissolve.

Earth Science Lab Manual

Name _____ Class _____ Date _____

3. **Using Models** Why must the ball of clay used in Part A be below the surface of the water in the bowl?

    The formation of caverns takes place at or below the water table in the zone of saturation.

4. **Forming Operational Definitions** The solution that you will use in Part B of this investigation is a saturated solution. Use *only* the information given in Part B to describe in your own words what *saturated* means.

    Using only the wording in the procedure, students should be able to infer that *saturated* refers to a solution in which the solvent (water) dissolves all of the solute (Epsom salts, borax) it can at that temperature.

5. **Inferring** In Part B of this investigation, why will you place the string into one of the cups for several minutes before setting up the equipment?

    The wet string will contain seed crystals that will increase the rate of formation of the dripstone features.

Materials (per group of four students)

~ 300 g modeling clay
5 small sugar crystals, cough drops, or hard candies
wooden skewer
medium-sized, clear, plastic bowl
hot tap water
250 mL white vinegar
large spoon
tongs
paper towels
serrated table knife

large graduated cylinder
large glass jar (1-L capacity)
800 mL Epsom salts, sugar, or table salt
60-cm cotton string
stirring rod
2 small metal washers
2 500-mL plastic drinking cups
all-purpose glue
piece of thick, corrugated cardboard (~ 20 cm × 40 cm)

Ask for volunteers to provide sugar crystals (*not* cubes), cough drops, or small hard candies for Part A. Note that the modeling clay for Part A must be insoluble in water.

For Part B, provide students with Epsom salts, sugar, and/or table salt to use to make the saturated solutions.

Prior to beginning this investigation, find an area where the setups can be placed so they will not be disturbed but can be easily observed and accessed.

Name _____ Class _____ Date _____

Safety 🬀🬁🬂🬃🬄🬅

Put on safety goggles and a lab apron. Be careful to avoid breakage when working with glassware. Be careful when handling sharp instruments. Never taste any substance in the lab unless instructed to do so. Use extreme care when working with heated equipment or materials to avoid burns. Wash your hands thoroughly after completing this investigation. Note all safety symbols next to the steps in the Procedure and review the meanings of each symbol by referring to the symbol guide on page xiii.

Procedure

Part A: Modeling Erosion by Groundwater

1. Put on safety goggles and a lab apron.
2. Flatten the modeling clay to form a circle with a diameter of approximately 10 cm. Place the sugar crystals, cough drops, or hard candies along one edge of the circle so that each touches at least one other crystal, cough drop, or hard candy.

   Remind students never to taste anything in the lab unless instructed to do so.

3. Shape the clay into a ball so that the crystals, cough drops, or hard candies are inside the ball. Place the ball of clay on your desk so that the solids are at the bottom of the ball.
4. Use the skewer to make at least five passageways that extend about three-fourths of the way into the ball of clay.
5. Place the bowl on your lab table or desktop. Fill the bowl halfway with hot tap water. Carefully add the vinegar to the water and stir the solution with the spoon.
6. Use the spoon to lower the clay ball into the bowl. If the clay is not completely covered with the water-vinegar solution, carefully add more hot water to the bowl until the clay is completely covered.
7. Allow the clay ball to stay in the solution for at least 30 minutes. Then proceed with Step 8. While you are waiting, start Part B of this investigation.
8. Layer two or three paper towels and fold them to form a square. Use the tongs to remove the clay from the solution and place it on the paper towels.
9. Use the serrated knife to cut the clay ball in half without squashing it. **CAUTION:** *Be careful when handling sharp instruments.*

10. In the space below, draw a cross-sectional, or side, view of the clay ball "cavern."

Cross Section of Model Cavern

Part B: Modeling Deposition by Groundwater

11. Use the graduated cylinder to measure and pour 650 mL of hot tap water into the large glass jar.

12. Add approximately half of the Epsom salts, sugar, or table salt to the water and stir until the crystals have completely dissolved.

13. Continue adding small amounts of the salts or sugar to the hot water until no more of the solid will dissolve.

14. Use the stirring rod to lower the string into the jar. Allow the string to soak in the solution for a few minutes.

15. Use the stirring rod to remove the string from the jar. Allow any excess solution to fall back into the beaker.

16. Tie a metal washer to each end of the string.

17. Place one end of the string into each cup, as shown in Figure 1. Allow some of the string to hang between the cups as shown.

18. With a partner, glue the cups onto the piece of cardboard as shown in Figure 1. Wait for the glue to dry then move the setup to a warm place where it will not be disturbed.

19. Use the graduated cylinder to measure and pour 300 mL of the saturated solution into each cup.

20. Wash your hands thoroughly after completing this investigation.

21. Observe your setup daily for about 2 weeks. If necessary, make and add more saturated solution to the cups.

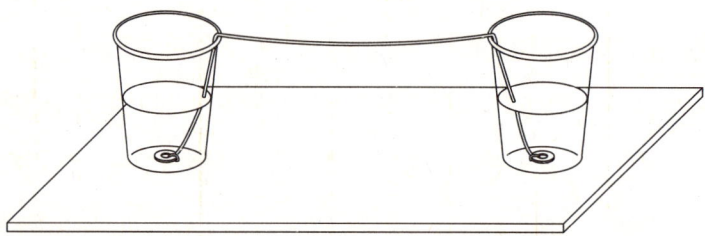

Figure 1

Name _____ Class _____ Date _____

Analysis and Conclusions

1. **Using Models** What did the clay and crystals, cough drops, or hard candy represent in Part A of this investigation?

   The clay represented rocks that are not soluble when exposed to acidic groundwater.

   The crystals, cough drops, or hard candies represented limestone.

2. **Inferring** What was the purpose of adding vinegar to the hot water in Part A of this investigation?

   Vinegar was added to simulate the acidity of groundwater.

3. **Relating Cause and Effect** Why did you make passageways into the ball of clay, and what do these features represent in the actual formation of caverns?

   The passageways allowed the water-vinegar solution to enter the clay and dissolve some of the hard substances within the clay. The passageways in the clay represented joints and bedding planes in actual rock layers.

4. **Using Models** Explain why the ball of clay in Part A of this investigation had to be completely submerged in the water-vinegar solution.

   Erosion forms caverns at or below the water table in the zone of saturation.

5. **Designing Experiments** Why could the string in Part B of this investigation not touch the sides of the cups?

   Had the string touched the sides of the cups, crystals would have formed on the insides of the cups rather than on the "ceiling" of the model cave.

6. **Observing** In Part B of this investigation, which formed first—the stalagmites or stalactites? Explain why.

   Stalactites formed first as water first rose up the string and then flowed down the string between the cups. As the water evaporated, small crystals formed first on the string. Eventually, as water dropped from the string, stalagmites formed on the cardboard.

Earth Science Lab Manual ▪ 63

7. **Applying Concepts** Explain why dripstone features form after the formation of the cavern in which they are found.

The deposition associated with the formation of dripstone features is not possible until the caverns are above the water table in the zone of aeration.

8. **Designing Experiments** How might you alter your procedures for both parts of this investigation to get better results?

Increasing the acidity of the "groundwater," increasing the number of passageways, and increasing water temperature are several ways to improve the formation of a model cavern (Part A). Modeling dripstone feature formation (Part B) is best done in areas with constant humidity and temperature. Also, increasing the temperature of the water allows saturation to occur much more quickly.

Going Further

Once your dripstone features have completely hardened, carefully remove them from the string and cardboard and compare and contrast the shape and size of your stalagmites and stalactites.
In most cases, the stalactites will be longer than the stalagmites. The stalactites also will be more conical than the stalagmites, which will be more massive in form. Some students may even have stalactites that are hollow.

Name _____ Class _____ Date _____

Chapter 7   Glaciers, Deserts, and Wind   Investigation 7

Continental Glaciers Change Earth's Topography

Have students review the information on topographic maps and glaciers in their textbooks.
SKILLS: Measuring, Calculating, Comparing and Contrasting, Interpreting Diagrams/Photographs
TIME REQUIRED: 30 minutes

Introduction

Glaciers are thick masses of ice that change Earth's surface. One type of glacier, **a continental glacier,** or ice sheet, is an enormous mass of ice that flows in all directions from one or more centers. Today, continental glaciers cover about 10 percent of Earth's landscape, where the climate is extremely cold. Thousands of years ago, however, these glaciers were more extensive than they are today. At one time, these thick sheets of ice covered all of Canada, portions of Alaska, and much of the northern United States. The impact that these ice sheets had on the landscape is still obvious today.

Landforms produced by continental ice sheets, especially those that covered portions of the United States, are mostly depositional in origin. Recall that there are two types of glacial deposits. **Stratified drift** is sediment that is sorted and deposited by glacial meltwater. **Till** is unsorted sediment deposited directly by a glacier. **Moraines** are ridges of till deposited as a glacier melts and recedes. **Ground moraines** are gently rolling plains of rocks and other glacial debris. Ground moraines can fill low spots and result in poorly drained swamplands. **End moraines** are deposits that form along the end of a melting glacier. An **outwash plain** is a ramp-like accumulation of sediment downstream from an end moraine. **Kettles** are glacial features that form when blocks of stagnant ice become buried and eventually melt. **Drumlins** are streamlined hills that are composed of till.

In this investigation, you will use a topographic map to examine some of the features produced by continental glaciation. Recall from Chapter 1 that a **topographic map** is a map that shows a bird's-eye, or top, view of an area. **Contour lines** on a topographic map connect points of equal elevation. The difference in elevation between adjacent contour lines is the **contour interval.** Closely spaced contour lines indicate a steep slope, while contour lines that are farther apart indicate a gentle slope. A **scale** shows how horizontal distances on the map are related to actual distances on Earth's surface.

Problem

How do continental glaciers change Earth's topography?

Pre-Lab Discussion

Read the entire investigation. Then work with a partner to answer the following questions.

1. **Measuring** What is the total length, in miles, of the glacial feature labeled **A**?

   The map scale is 1 inch equals 1 mile. The glacial feature labeled **A** is approximately 8.5 miles long.

2. **Calculating** Approximately how long, in miles, is Blue Spring Lake? Show your calculations.

   The map scale is 1 inch equals 1 mile. The length of the lake on the map is 13/16 of an inch.

   5280 ft/mi × 13/16 mi = 4209 ft

   4290 ft/5280 ft = 0.8 mi

3. **Interpreting Diagrams/Photographs** How does the topography in the southeast corner of the map compare with the topography in the northwest part of the map?

   The topography in the southeast corner of the map has a higher elevation, and it is more irregular than the land in the northwest.

4. **Interpreting Diagrams/Photographs** What features on the map indicate that portions of the area are poorly drained? Where are these features located?

   Marshes and swamps in the central and northwest portions of the map indicate that the area is poorly drained.

5. **Comparing and Contrasting** How are glacial drift and till alike? How are they different?

   Glacial drift and till are sediment deposited by glaciers. Drift is sorted sediment deposited by meltwater. Till is unsorted debris that is deposited directly from the ice.

Name _____ Class _____ Date _____

Materials *(per pair of students)*

metric ruler
Resource 1 in the DataBank
Resource 9 in the DataBank
calculator (optional)

You may want to provide students with copies of the Whitewater topographic map, Resource 9, in the DataBank.

Procedure

1. Closely examine the Whitewater, Wisconsin, topographic map (Resource 9 in the DataBank). If necessary, refer to the map symbol guide (Resource 1 in the DataBank).
2. Use the topographic map and what you have learned about glaciers to answer the **Analysis and Conclusions** questions.

Analysis and Conclusions

1. **Interpreting Diagrams/Photographs** The feature labeled **A** on the map is unsorted glacial debris. What is this structure, and how did it form?

   The feature labeled A is a moraine that formed as the leading edge of the glacier melted and receded.

2. **Interpreting Diagrams/Photographs** What are the structures on the map labeled **B**? Find and label another example of this structure on the map.

   The structures labeled B are drumlins. Another drumlin, which is much larger than those labeled, is located southwest of Zion Cemetery.

3. **Inferring** Where on the map is the likely location of the outwash plain?

   The southeast quarter of the map is the outwash plain.

4. **Interpreting Diagrams/Photographs** What are the structures on the map that are labeled **C**? How do these structures form?

   The structures labeled C are kettles, or kettle lakes. They form when blocks of stagnant ice become buried and eventually melt.

5. **Inferring** Where on the map might you find ground moraine?

   Ground moraine was deposited in the western and northeastern parts of the area shown.

6. **Applying Concepts** In which direction did the continental glacier that changed this part of Wisconsin move? Give two lines of evidence to support your answer.

The ice moved from north to south or slightly southeast. This direction of movement is supported by the orientation of the drumlins, whose steep sides face the direction from which the ice advanced and the position of the moraine labeled **A**.

Go Further

If necessary, review how to make a topographic profile in Investigation 6A, Rivers Shape the Land. Then use the Whitewater topographic map to make a northwest-southeast topographic profile from the Scuppernong River to the city of Little Prairie.

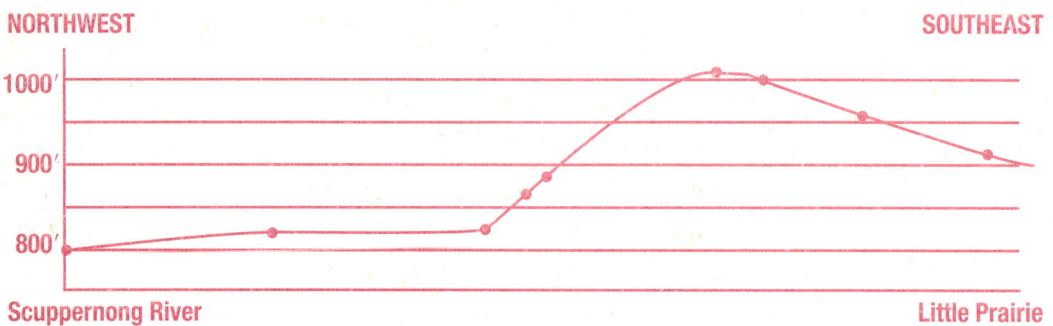

Name _____ Class _____ Date _____

Chapter 8   Earthquakes and Earth's Interior   Investigation 8A

Modeling Liquefaction

Have students review the information on destruction from earthquakes in their textbooks.
SKILLS: Observing, Comparing and Contrasting, Using Models   **TIME REQUIRED:** 30 minutes

Introduction

When coffee is packed in a vacuum, the air is removed from the package. This vacuum packaging causes the coffee grounds to compact. When the package is opened, air moves among the coffee grounds and allows them to once again move freely within the package.

Water trapped in sandy soil can cause the soil to behave like a liquid if external forces are applied faster than the water can escape the soil. The soil behaves like a liquid when water pressure is equal to the forces acting on the soil. This process is called **liquefaction.** Liquefaction often occurs when an earthquake strikes because the forces associated with the shaking and vibration at the surface of Earth cause sand particles to move away from one another. Because the grains are not in contact with one another, the soil loses strength and is said to liquefy.

In this investigation, you will model sand liquefaction, a common phenomenon in certain earthquake-prone areas.

Problem

How can you model sand liquefaction?

Pre-Lab Discussion

Read the entire investigation. Then work with a partner to answer the following questions.

1. **Posing Questions** Write a question that summarizes the purpose of this investigation.

   Sample questions: How does wet sand react during an earthquake? Do wet and dry sand react

   differently when outside forces are applied to them?

2. **Controlling Variables** What is the independent variable in this investigation?

   The independent variable in this investigation is water added to the sand.

3. **Controlling Variables** What is the dependent variable in this investigation?

   The dependent variable is the movement of the sand when a force (squeezing) is applied.

Earth Science Lab Manual ▪ 69

Name _____ Class _____ Date _____

4. **Designing Experiments** Why will you glue filter paper to the bottom of the rubber stopper?

   The filter paper will prevent sand from being sucked out of the balloon, and it helps to create a better vacuum.

5. **Comparing and Contrasting** How do you think the balloon holding dry sand will feel when you squeeze it? How will this compare with the balloon holding wet sand?

   The dry sand will feel more gritty than the wet sand, and the wet sand will move more easily than the dry sand.

Materials *(per pair of students)*

large, round rubber balloon
large drinking straw
rubber stopper with hole for drinking straw
filter paper
scissors
all-purpose glue
funnel
500 g clean, dry, medium-grained sand
250-mL graduated cylinder
tap water
measuring cup
vacuum pump

Play sand or aquarium sand works best for this activity.

Make sure that the holes in the stoppers are just large enough for the straws. If the holes are too large, students will have problems creating a partial vacuum within the balloons.

Safety

Put on safety goggles. Be careful to avoid breakage when working with glassware. Be careful when handling sharp instruments. Observe proper laboratory procedures when using electrical equipment. Use the vacuum pump to remove most of the air from the balloon. Note all safety symbols next to the steps in the Procedure and review the meanings of each symbol by referring to the symbol guide on page xiii.

Procedure

1. Put on safety goggles.
2. Insert the straw into the hole in the rubber stopper so that the bottom edge of the straw is flush with the bottom of the stopper.
3. Trace the circumference of the bottom of the stopper onto the filter paper. Cut out the circle and glue it to the bottom of the stopper. Wait for the glue to dry completely.

Name _____ Class _____ Date _____

4. Work with a partner and use the funnel to fill the balloon halfway with sand.

5. Hold the balloon by the neck and gently squeeze the sand-filled portion with your fingers. Record your observations in the Data Table.

6. Pull the neck of the balloon over the bottom of the rubber stopper. Attach the tubing from the vacuum pump to the top of the straw and remove nearly all of the air from the balloon. Pinch the neck to close off the balloon.

If necessary, demonstrate how to use the vacuum pump. Remind students that they should not remove too much air from the balloons.

7. Again, hold the balloon by the neck and gently squeeze the sand-filled portion with your fingers. Record your observations in the Data Table.

8. Remove the tubing from the straw and remove the balloon from the rubber stopper.

9. Use the graduated cylinder and funnel to add water to the sand-filled balloon. Add only enough water to make the sand wet. The amount of water you put into the balloon should be equal to about half of the amount of sand in the balloon.

10. Gently squeeze the balloon holding the sand-water mixture to wet the sand grains evenly.

11. Gently force excess air out of the balloon and tie the balloon at the bottom of the neck.

12. Gently squeeze the balloon and record your observations in the Data Table.

13. Gently but quickly squeeze the balloon five or six times. Record your observations in the Data Table.

Observations

DATA TABLE

Trial	Observations
Balloon with sand	At this point, the sand still contains air pockets, so it is free to move when pressure is applied to the balloon.
Balloon with sand (air removed)	The sand in the balloon will feel harder or denser than it did before the air was removed.
Balloon with sand and water (slow squeezing)	The sand in the balloon moves easily, but one is still able to feel the grains.
Balloon with sand and water (rapid squeezing)	The sand in the balloon moves easily, but it is difficult to feel the sand grains.

Name _____ Class _____ Date _____

Analysis and Conclusions

1. **Comparing and Contrasting** Compare and contrast the ease of movement of the dry sand in the balloon before and after air was removed from the balloon. Explain why this happened.

   The dry sand moved more freely before the air was removed because the sand grains were farther apart. There was more air between adjacent sand grains, allowing more movement of grains before the air was removed. After the air was removed, there was less air between the grains, making it difficult for the sand to move.

2. **Comparing and Contrasting** Compare and contrast the movement and feel of the wet sand when you squeezed slowly and when you squeezed rapidly.

   The sand felt gritty when it was slowly squeezed. When it was rapidly squeezed, it felt more like a liquid.

3. **Inferring** What role does friction play in liquefaction?

   Sand grains that are closer together have a higher frictional force to overcome in order to move the grains.

4. **Using Models** What happens when water pressure increases in sandy soil in an earthquake-prone area?

   As water pressure increases, sand grains move more freely, which causes the sand to liquefy. The resulting fluid-like substance often collapses during an earthquake, causing much damage.

5. **Applying Concepts** Study the observations you recorded in the Data Table. If you were to build a structure on one of these soils, which would you choose to prevent earthquake damage to the structure? Explain your choice.

   The dry sand with air removed would be best. It would minimize earthquake damage because the sand grains are close together and move little when outside forces are exerted on them.

6. **Designing Experiments** How could you redesign this investigation to produce better results?

   Sealing the hole in the stopper with clay or some other material would help create a better vacuum. Using a stronger pump to remove air from the balloon also would improve the procedure followed in this investigation.

Name _____ Class _____ Date _____

Chapter 8   Earthquakes and Earth's Interior Investigation 8B

Design and Build a Simple Seismograph

Have students review the information on seismographs in their textbooks. You may want students to do some additional research on seismographs and how they work before assigning this investigation. **SKILLS:** Designing Experiments, Making Judgments, Measuring, Using Models **TIME REQUIRED:** three 45-minute periods—one period to propose a design, one period to construct the device, and one period to test the device and modify it, if necessary

Introduction

A **seismograph** is an instrument that records movements of the ground caused by earthquakes, explosions, and other ground-shaking events. All seismographs have a weight, a support that is anchored to the ground, and a device that records the vibrations.

When earthquake waves reach a seismograph, the inertia of the suspended weight keeps it stationary while the support vibrates with the ground motion. Some seismographs record vertical motions. The weight on such an instrument is often suspended from a metal spring, and the movements are recorded on a vertical drum.

Other seismographs record horizontal movements of the ground. These instruments use a weight suspended from a wire to sense and record vibrations on a horizontal drum. A record of any ground movements is called a **seismogram.**

In this investigation, you will design, build, and test a simple seismograph to record movements. You will also evaluate the seismograms produced by your instrument.

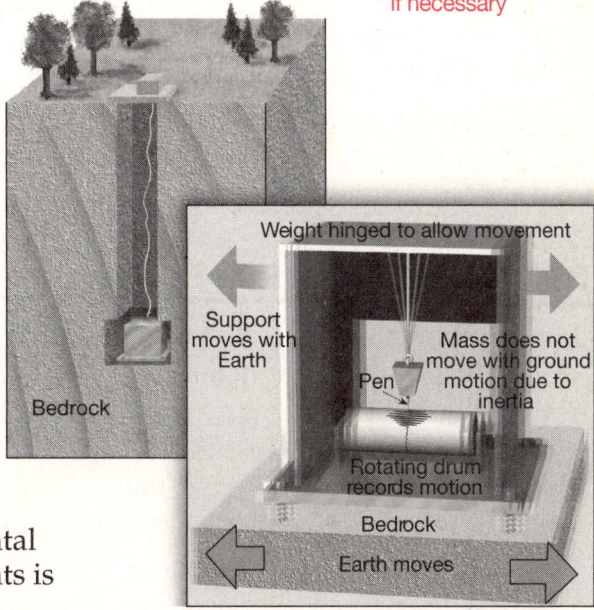

Figure 1 How a Seismograph Works
The inertia of the suspended mass tends to keep it motionless, while the recording drum, which is anchored to bedrock, vibrates in response to seismic waves.

Problem

How can you build and test an instrument that will record ground vibrations?

Pre-Lab Discussion

Read the entire investigation. Then work with at least three other students to answer the following questions.

1. Relating Cause and Effect Explain how a seismograph works.

The instrument includes a suspended weight that is stationary as the result of inertia, and a

support, or base, that is anchored to the ground. When the ground moves, the weight remains

motionless while the base of the instrument moves with the ground. A pen records vibrations

on a rotating drum.

Earth Science Lab Manual ▪ 73

Name _____ Class _____ Date _____

2. **Inferring** Look again at the suggested materials list. Why do you think one of the suggested materials is a small brick?

   The brick acts as a weight.

3. **Using Models** Why do you have to make sure that the roll of adding machine paper is able to turn?

   The pencil or other recording device will keep marking successive vibrations in the same place on the paper if the roll does not turn.

4. **Relating Cause and Effect** How will your group produce "earthquakes" of various magnitudes?

   The "earthquake magnitude" can be varied by shaking the table with varying forces.

Suggested Materials *(per group of four or five students)*

thin (~2.5 cm thick) wood planks	duct tape	small (~30–40 cm long) wood beams
heavy cord or wire	hammer	vice grips
hand saws	sharp pencil with soft lead	nails and screws
wire cutters	wood dowels	metric ruler
hand drills	adding machine tape	small, thin brick

Note: Ask your teacher for any other materials that you think might be useful to build your seismograph.

Safety

Wear your safety goggles during this entire investigation. Use care when handling sharp instruments. Observe proper safety procedures when using the hammer, saws, and drills. Be careful when handling the wood to avoid getting splinters in your skin. Note all safety symbols next to the steps in the Procedure and review the meanings of each symbol by referring to the symbol guide on page xiii.

Design Your Own Investigation

1. Put on your safety goggles.
2. Reread the **Introduction** and the list of **Suggested Materials.** Also refer to Figure 1, which illustrates how a seismograph works, and Figures 2 and 3, which show examples of built seismographs. Use this information to draw and label a sketch of your proposed seismograph in the space provided on page 76. Make any necessary notes on your design.

Students must work cooperatively to design and build the instruments. Have each group determine how each student in the group will participate in this investigation.

Remind students to use care when working with saws and drills. Demonstrate the proper way to use these tools.

If your classroom does not have the space needed to allow students to cut the wood, you might ask adult volunteers or the woodshop teacher to help you precut the pieces of wood needed to build the instruments.

Name _____ Class _____ Date _____

3. On page 77, write the steps you will take to build your seismograph. Add as many steps as you need. Be sure to include any additional materials you may need.

4. Have your teacher approve your design and procedure. Work with at least three other students to construct your seismograph. **CAUTION:** *Use care when using saws and drills. Also take care when using the hammer and sharp instruments.*

Closely supervise students as they build their instruments.

5. When your instrument is complete, test it to make sure that it records vibrations. If necessary, determine how to change your instrument so that it works properly. Discuss your proposed changes with your teacher and have your teacher approve the changes. Make the necessary changes and test your seismograph again. Continue to make changes as necessary.

6. Once your seismograph is working properly, anchor it to a tabletop with vise grips. Then simulate an earthquake 10 times. Each "earthquake" should be of various magnitudes. Be sure that the paper is being pulled at a regular rate as the simulated quake strikes. Label each of the seismograms accordingly.

Students should simulate the earthquakes by shaking the table enough to produce noticeable marks on the paper, but not so hard that the table becomes unstable.

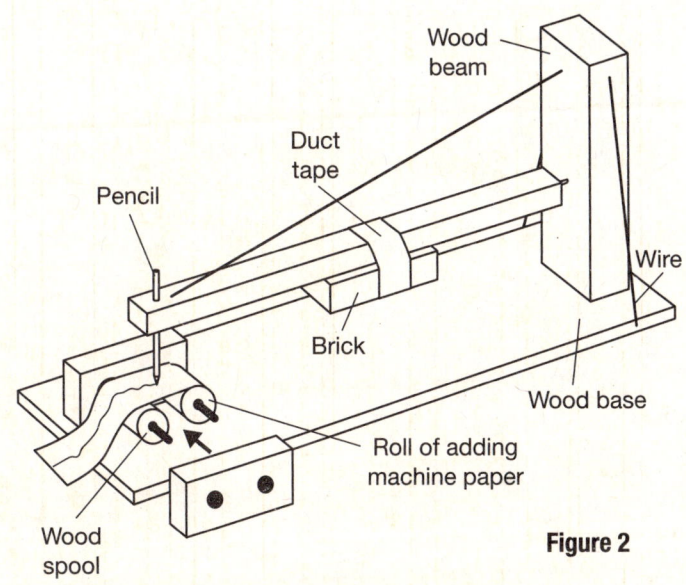

Figure 2

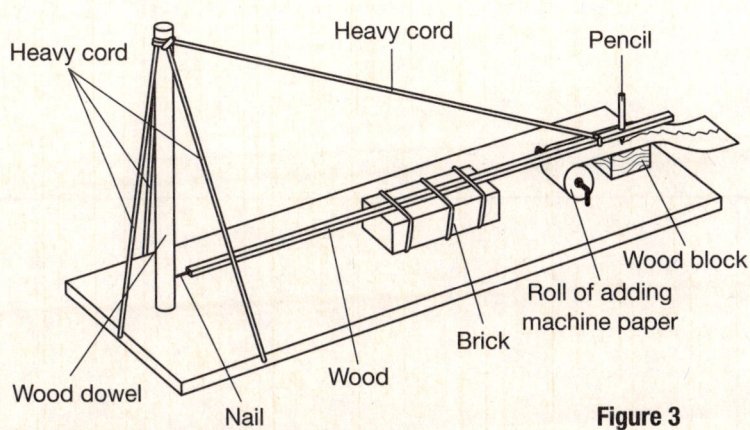

Figure 3

Earth Science Lab Manual ▪ 75

Name _____ Class _____ Date _____

Labeled Sketch of Proposed Seismograph

Name _____ Class _____ Date _____

Procedure

Step 1: _____
_____
_____
_____
_____

Step 2: _____
_____
_____
_____
_____

Step 3: _____
_____
_____
_____
_____

Step 4: _____
_____
_____
_____
_____

Step 5: _____
_____
_____
_____
_____

Step 6: _____
_____
_____
_____
_____

Name _____ Class _____ Date _____

Analysis and Conclusions

1. **Observing** What type of movement—horizontal or vertical—were you able to record with your seismograph?

   Students were able to record horizontal movements with their instruments.

2. **Using Models** What is the purpose of the brick?

   The brick allows the arm of the instrument to stay in place so that only the frame will vibrate during the simulated quakes.

3. **Using Models** Why did you anchor the seismograph to a tabletop before simulating the earthquakes?

   In order for a seismograph to record vibrations, some point on the instrument—the suspended mass or pendulum—must remain at rest. The fixed frame of a seismograph must be fastened to the ground (tabletop) so that it, not the suspended mass, moves during an earthquake.

4. **Comparing and Contrasting** How did the seismograms produced during each "earthquake" compare? How were they different?

   Each seismogram is a record of the horizontal movement of the simulated earthquake. The stronger "earthquakes" produced lines with greater amplitudes than the weaker "earthquakes."

5. **Designing Experiments** How could you redesign this investigation to produce better results?

   The instrument made in this investigation only records one direction of horizontal movement. To record other horizontal movements produced by the quakes, students would need another instrument oriented perpendicular to the first seismograph. Attaching a motor to automatically turn the paper also would improve the seismograms produced.

Go Further

Draw and label a seismograph that could be built to measure vertical movements.

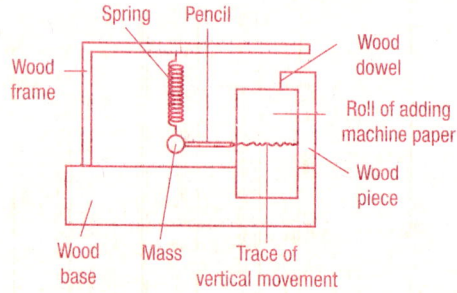

Earth Science Lab Manual • 78

Name _____ Class _____ Date _____

Chapter 9   Plate Tectonics   Investigation 9

Modeling a Plate Boundary

Introduction Have students review the information on earthquake patterns in their textbooks.
SKILLS FOCUS: Using Graphs, Using Models **TIME REQUIRED:** 45 minutes

The lithosphere is divided into moving segments called **plates.** The plates move as units relative to all other plates. All major interactions occur among individual plates along boundaries. Scientists first attempted to outline the plate boundaries by using locations of earthquakes. Later research showed plates bounded by three distinct types of boundaries, which exhibit different types of movement.

A **convergent boundary** is formed when two plates slowly move together. At this boundary, the leading edge of one plate is bent downward, sliding beneath the second plate. This process is called subduction, and the convergent boundaries are called **subduction zones.** The surface expression produced by one plate sliding below another plate is an **ocean trench.**

Just south of the Aleutian Islands in the northern Pacific Ocean, the Pacific plate moves northward and is subducted beneath the North American plate. A large number of earthquakes occur in this region. In this investigation, you will use earthquake data from one part of this region to form a model of the convergent boundary between the two plates.

Problem

How can you use earthquake data to model a convergent boundary between two plates?

Pre-Lab Discussion

Read the entire investigation. Then work with a partner to answer the following questions.

1. **Posing Questions** Write a question that summarizes the purpose of this investigation.

   Sample question: How can you use earthquake data to model a convergent boundary and

   estimate the location of an ocean trench?

   _____

2. **Controlling Variables** What is the dependent variable in this investigation?

   The dependent variable is the depth of the foci of the earthquakes.

   _____
   _____

Earth Science Lab Manual ▪ 79

Name _____ Class _____ Date _____

3. **Controlling Variables** What is the independent variable in this investigation?

   The independent variable is the latitude of the earthquakes. The latitude of an earthquake is related to the distance from the ocean trench.

4. **Inferring** Why should the graph of earthquake depth vs. earthquake latitude have a zero at the top of the vertical axis?

   Graphing in this way resembles the physical system. The data points move downward as depth increases.

5. **Predicting** How do you think earthquake depth is related to the distance from an ocean trench?

   Answers will vary, but most students should predict that shallow-focus earthquakes tend to occur near the ocean trench. The depth of the focus increases as the distance from the trench increases.

Materials (per pair of students)
ruler
protractor
Resource 3 in the DataBank

Procedure
1. Examine the map on Resource 3 in the DataBank. Study the convergence of the Pacific plate and the North American plate just south of the Aleutian Arc of volcanic islands in the northern Pacific Ocean.

2. Draw and label a diagram showing how the edges of the Pacific plate and the North American plate converge. Use Figure 1 to help you draw this diagram.

Diagrams should provide a simplified version of Figure 1. The upper plate should be labeled as the North American plate, and the descending plate should be labeled the Pacific plate.

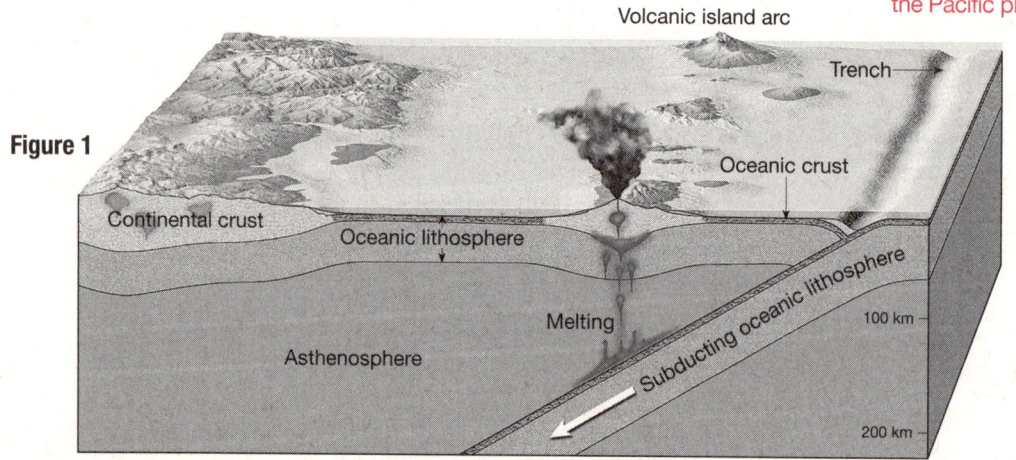

Figure 1

Name _____ Class _____ Date _____

3. The Data Table below shows the depths of foci and latitudes of earthquakes in the Aleutian Islands. All of the earthquakes in the table occurred near 180°W longitude. Examine the table of earthquake data and record any patterns you observe.

Answers will vary, but most students should note that the higher latitudes correspond to greater depths of the foci.

DATA TABLE

Earthquake	Year	Latitude of Epicenter (°N)	Depth of Focus (km)
1	1982	51.39	51
2	1983	51.97	116
3	1984	51.13	15
4	1985	52.36	213
5	1985	52.62	233
6	1986	51.70	67
7	1986	52.31	170
8	1987	51.29	22
9	1987	51.93	94
10	1990	52.30	143
11	1991	51.96	108
12	1992	52.01	99
13	1992	52.13	130
14	1992	52.48	211
15	1995	51.19	29
16	1997	51.28	33
17	1998	51.59	43
18	1999	51.87	72
19	2000	51.60	71
20	2001	51.32	55
21	2001	51.77	79
22	2003	51.15	11
23	2003	52.13	180

4. Use the information in the Data Table to construct a graph showing the location and depth of the earthquakes. Use the following grid and plot the latitude on the horizontal axis and the depth of the focus on the vertical axis. Number the vertical axis with zero at the top and maximum depth at the bottom. Give your graph an appropriate title.

5. After plotting the data, draw a straight line that comes as close as possible to each of the data points.

Name _____ Class _____ Date _____

Title: _____

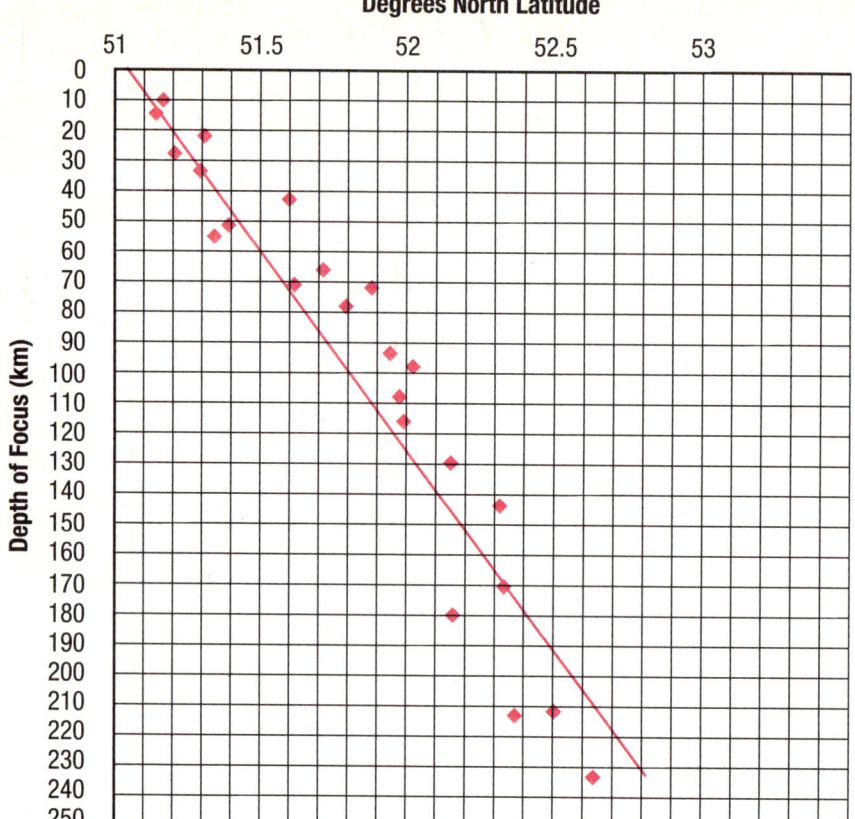

Analysis and Conclusions

1. **Using Graphs** What relationship exists between the depths of the earthquake foci and the latitude?

   The depths of the foci increase as the latitudes increase from about 51.1°N.

2. **Analyzing Data** How does the graph illustrate that the boundary between the Pacific plate and the North American plate is a convergent boundary?

   The earthquakes tend to occur along the boundary between the upper plate and the descending plate. This boundary becomes deeper as the latitude increases, confirming that the Pacific plate is descending beneath the North American plate.

3. **Applying Concepts** The Aleutian trench is located where the two plates meet at the surface of the lithosphere. Use the graph to determine the approximate latitude of the Aleutian trench at 180°W longitude. Explain your answer.

   Most earthquakes occur at the convergent boundary. At the location of the trench, this boundary is at the surface and its depth is zero. Therefore the trench is located where the line through the data intersects with the line $y = 0$. This occurs at approximately 51.1°N latitude.

Earth Science Lab Manual

Name _____ Class _____ Date _____

4. **Using Graphs** Use the slope of the graph to determine how quickly the convergent boundary descends as latitude increases.

   Answers will vary, but students should find that the convergent boundary descends approximately 140 km per a one-degree increase in latitude.

5. **Calculating** A change in latitude of one degree corresponds to a distance of approximately 111 km along a north–south line. Using this information and your answer to Question 4, determine how far the convergent boundary descends as it moves 1 km northward. Show your work.

   Answers will depend on students' answers to Question 4. Sample answer:
   $$\frac{140 \text{ km}}{1 \text{ degree}} = \frac{140 \text{ km}}{111 \text{ km}} = \frac{1.26 \text{ km}}{\text{km}},$$ so the boundary descends about 1.26 km for each km northward.

Use your results from Question 5 to help you answer Questions 6 and 7.

6. **Using Models** Draw a triangle that shows a side view of the convergent boundary with the correct scale relationship between horizontal distance and depth. Measure the angle at which the subducted plate descends beneath the upper plate. This is called the subduction angle.

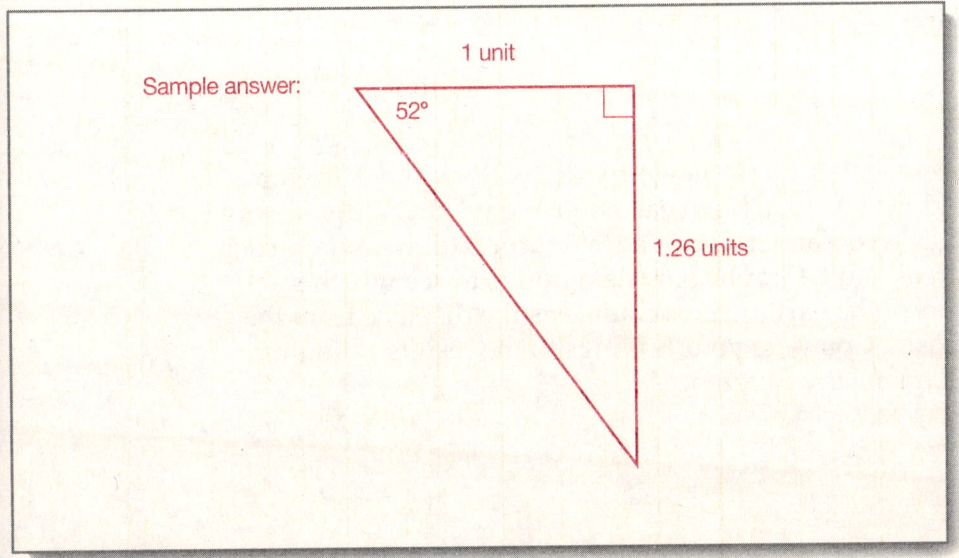

Sample answer: (triangle with 1 unit horizontal, 1.26 units vertical, 52° angle)

Earth Science Lab Manual ▪ 83

Name _____ Class _____ Date _____

7. **Calculating** Imagine there is a volcanic island arc on the surface of the lithosphere above the area where the descending plate reaches a depth of 100 km. Approximately how far north of the trench are the islands located? Show your work.

> Answers will depend on students' answers to Question 5. Sample answer: The depth of the descending plate increases 1.26 km for each 1 km northward. Therefore, use the ratio
>
> $$\frac{1.26 \text{ km}}{1 \text{ km northward}} = \frac{100 \text{ km}}{x \text{ km northward}}.$$
>
> $x = \frac{100}{1.26}$ km = 79.4 km, so the island arc should be located about 80 km north of the trench.

8. **Evaluating and Revising** Are all of the data points you plotted on the graph close to the straight line you drew to best fit the data? Provide possible explanations for your answer.

   No, the data are somewhat scattered. There may have been experimental errors in determining the depths of foci or the locations of the epicenters, or perhaps some of the earthquakes did not occur along the plate boundary. Also, the straight-line model of the plate boundary is only an approximation. The actual boundary probably is irregular or has some curvature.

9. **Applying Concepts** Is the year in which an earthquake occurred an important variable in this investigation? Explain.

   No. Plates typically move only a few centimeters per year relative to each other, so the locations of the plates are practically unchanged over two decades.

Go Further

The United States Geological Survey provides earthquake data from locations around the world. Obtain earthquake data near a divergent boundary such as an oceanic ridge or a transform fault boundary such as the San Andreas Fault. Graph some data points to see whether the depths of the foci of the earthquakes change as the distance from the boundary increases. Compare your findings to the results from this investigation. Explain any differences.

The data and information students need for this activity can be found on

For: Chapter 9 Resources
Visit: PHSchool.com
Web Code: cjk-9999

Other types of boundaries do not exhibit the same relationship between depth of focus and distance. There is an absence of deep-focus earthquakes along divergent boundaries. These findings are consistent with the plate tectonics model.

Earth Science Lab Manual ▪ 84

Name _____ Class _____ Date _____

Chapter 11   Mountain Building

Interpreting a Geologic Map

Investigation 11

Have students review the information on topographic and geologic maps and folds and faults in their textbooks. **SKILLS FOCUS:** Observing, Measuring, Applying Concepts, Interpreting Diagrams **TIME REQUIRED:** 35 minutes

Introduction

Geologic maps show the distribution of rock units at Earth's surface, as if the soil and other loose material had been stripped away. The rock is divided into units called **formations** that can be recognized and traced across the map area. Formations are identified by a color and/or pattern. Each unit is labeled with a letter abbreviation that indicates the age and name of the unit. The boundaries between formations are indicated by solid lines called **contacts.** If the boundary is uncertain or difficult to accurately locate, the contact will be shown as a dashed line. Geologic maps also show features such as **faults, folds,** and **unconformities,** as well as contour lines that indicate the elevations of the formations.

In this investigation, you will identify and interpret features on a geologic map for a portion of the northern Rocky Mountains.

Problem

What information can be obtained from a geologic map?

Pre-Lab Discussion

Read the entire investigation. Then work with a partner to answer the following questions.

1. **Posing Questions** Write a question that summarizes the purpose of this investigation.

   Sample question: How are geologic maps useful?

   _____

   _____

2. **Interpreting Diagrams** What do the lightly colored lines that are labeled with numbers represent?

   The lines are contour lines that represent lines of equal elevation and show the topography.

   _____

   _____

3. **Observing** How can you determine the age of the rock units shown on the map?

   The age of the rock units can be determined by using the map key or explanation.

   _____

   _____

Earth Science Lab Manual ▪ 85

Name _____ Class _____ Date _____

Materials *(per group of students)*
Resource 6 in the DataBank
Resource 7 in the DataBank
map of North America or atlas
colored pencils
string
ruler
protractor

Students can use string to mark off distances on the map. They can then lay the string next to the ruler to determine the distance. Give each group of students a length of string approximately 30 cm long. The map is not in SI units. All United States Geological Survey maps are in English units, although some newer maps use both units. However, the topography will always be shown in feet. If rulers with English units are not available, students can use the string and the bar scale on the map to calculate distances. Have students study the map and map key closely before beginning the investigation.

Procedure

1. Carefully study the geologic map, Resource 6. Match rock units, shown on the map as areas of different colors and patterns, with the descriptions of those units on the map key.

2. The square grids on the topographic base map represent numbered sections in a Township and Range grid system. The grids can be used to help locate features on the map. Determine the state, counties, latitude and longitude of the geologic map using the map, map key, and a map of North America or an atlas.

   The map is of Jefferson and Broadwater counties in Montana at 46°N latitude and 112°W longitude.

3. To measure distances on the map that are not on a straight line, use a string to follow the outline of the feature. Curve the string along the feature, holding one end of the string where you want to begin the measurement. Hold the string at the end of the feature. Be sure to continue holding the string at the starting point and ending point. Straighten out the string along the bar scale on the map or along the ruler to determine the measurement of the feature.

4. Maps provide a two-dimensional picture of the geology. In order to examine the geology in the third dimension, a cross section or profile is needed. To construct a cross section, first determine the line of section. A line of section XX' has been provided on the map. First, take a piece of paper and lay it on the map along line XX'.

5. Make small marks on the paper where a geologic contact or fault crosses the line of section. Transfer these contacts to the graph where you will construct your cross section.

6. Examine the map to determine what formations are present in your cross section. Label the formations on your cross section.

7. To determine the orientation of the geologic contacts, examine the map for strike-and-dip symbols. These are small T-shaped symbols with a number next to the short end of the *T*. This number represents the angle the rock units make with a horizontal line. This angle is called the **dip** of the unit. Locate any strike-and-dip symbols near the line of section.

Earth Science Lab Manual • 86

Name _____ Class _____ Date _____

8. Using the protractor, draw the contacts on your cross section with the correct angles.

9. On your cross section, draw all the contacts as straight lines from the angle. Label all the formations using the labels on the map key in the DataBank.

10. Make a cross-section key using the same labels and colors found on the map key in the DataBank. Use colored pencils to complete your cross section.

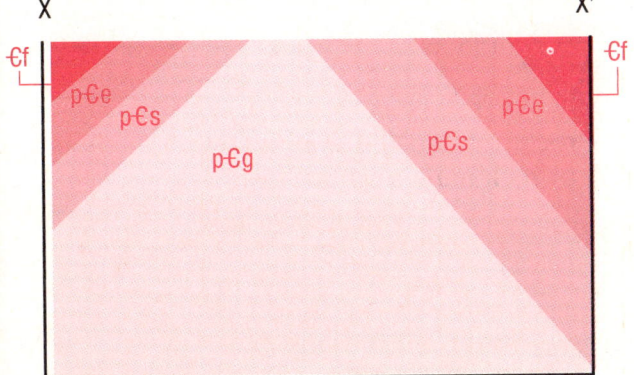
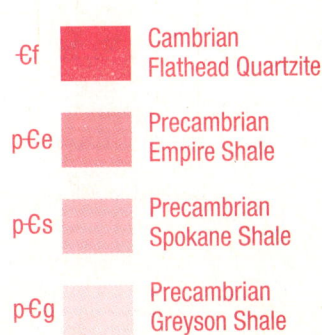

Cross Section XX'

Analysis and Conclusions

1. **Observing** What is the scale of the map?

   The scale of the map is 1 inch = 1 mile.

2. **Measuring** What is the length of the stream on the west side of the map from the north edge of section 10 to the west edge of section 15?

   The stream is approximately 4.25 miles long between the section lines indicated.

3. **Observing** What is the contour interval on the map?

   The contour interval is 40 feet.

4. **Observing** In what direction does the stream in Question 2 flow? Explain how you can determine the direction of flow.

   The stream flows southward with a few curves to the west. The direction of the stream flow can

   be determined by examining the contour lines because the stream flows from a higher elevation

   toward a lower elevation.

5. **Calculating** What is the gradient of the stream from the north edge of section 10 to the west edge of section 15? **NOTE:** Gradient is the change in elevation divided by the change in distance.

   $$6320 - \frac{5670 \text{ ft}}{4.25 \text{ mi}} = \frac{650 \text{ ft}}{4.25 \text{ mi}} = \text{about 153 ft/mi}$$

Earth Science Lab Manual ▪ 87

6. **Analyzing Data** What geologic structure is exposed on the eastern half of the map? Explain.

   The structure is a fold. It is a type of fold called an anticline. The oldest rock units are at the center of the fold, and the rock units dip away from the center in an up-fold or anticline.

7. **Interpreting Diagrams** What is the oldest rock unit that is exposed on the map?

   The Precambrian Greyson shale is the oldest rock unit exposed on the map.

8. **Interpreting Diagrams** What age is the rock unit labeled *ad* in the southeast corner of the map? What type of rock makes up this unit?

   The rock unit labeled *ad* is Cretaceous in age, and it is composed of igneous intrusive rocks.

9. **Drawing Conclusions** If you walk across a fault from the side where the Flathead quartzite is exposed to the side where the Empire shale is exposed, which side of the fault has moved up? Explain how you can determine which side of the fault moved.

   The side of the fault where the Empire shale is exposed has moved up relative to the side of the fault where the Flathead quartzite is exposed. The Empire shale is Precambrian in age, so it is older than the Cambrian Flathead quartzite. The older Precambrian rock unit was originally deposited first beneath the younger Cambrian rocks. The side of the fault where the older rocks are exposed moved up relative to the younger rocks.

Go Further

Construct a cross section that extends across the entire map by extending the XX' line of section to the western edge of the map. What structure is exposed in the northwest corner of the map?

The extended cross section should show a syncline in the northwest corner of the map with an anticline on the eastern half of the map, as shown in the first cross section.

Name _____ Class _____ Date _____

Chapter 12   Geologic Time   **Investigation 12**

Modeling Radioactive Decay

*Refer students to their textbooks for a discussion of dating objects using radioactivity. Students will find the Half-Life Decay Curve helpful when answering the Analysis and Conclusions questions. **SKILLS FOCUS:** Using Models, Observing, Measuring, Using Tables and Graphs **TIME REQUIRED:** 40 minutes*

Introduction

When scientists learned to measure radioactive decay, they gained the ability to determine the ages of many rocks, minerals, fossils, and archaeological objects. **Radiometric dating** is the name of the procedure that scientists use for these age determinations. It relies on the constant rate of radioactive decay that occurs among radioactive isotopes such as uranium-238, thorium-232, potassium-40, and carbon-14. This rate is expressed as a **half-life,** which is the amount of time it takes for one-half of the nuclei in a sample of a radioactive isotope to decay into the stable daughter product.

   In this investigation, you will model radioactive decay using pennies and then use your results to practice the radiometric dating procedure.

Problem

How can you model radioactive decay using pennies?

Pre-Lab Discussion

Read the entire investigation. Then work with a partner to answer the following questions.

1. **Using Models** What is the advantage of creating a simple model of radioactive decay?
   Creating a model of radioactive decay makes the concept easier to understand.

2. **Inferring** Why is a penny useful for representing a radioactive isotope?
   Just as a radioactive isotope has a 50:50 chance of decaying during one half-life, a tossed penny has a 50:50 chance of landing heads-up.

3. **Using Models** What represents the parent atoms in this activity? What represents the daughter atoms?
   The heads-up pennies represent the parent atoms. The tails-up pennies represent the daughter atoms.

Earth Science Lab Manual ▪ **89**

Name _____ Class _____ Date _____

4. Predicting How will the abundance of the heads-up and tails-up pennies change over time during this activity?

The heads-up pennies will decrease in abundance. The tails-up pennies will increase in abundance.

5. Contrasting Identify three ways in which this model differs from the actual process of radioactive decay.

Answers will vary, but students should note that the time frame is far shorter; measurements are simpler; materials are safer; the numbers are smaller; the amount of radioactive isotope will never reach zero, while the number of heads-up pennies will reach zero.

Materials *(per pair of students)*
flat box with a lid, such as a shoebox
100 pennies

The transparent plastic boxes used in topographic mapping activities also work well for this activity. Otherwise, have students begin collecting shoeboxes well in advance of the activity. M&Ms can be substituted for pennies.

Procedure
1. Place 100 pennies heads-up in the bottom of the box.
2. Cover the box with its lid and shake the box vigorously.
3. Set the box down and open it. Remove all pennies that are tails up. Count the remaining pennies and record the number in the Data Table below.
4. Cover the box containing the remaining heads-up pennies. Shake the box vigorously.
5. Repeat Steps 3 and 4 until no pennies remain in the box.

Observations

DATA TABLE

Shake #	Number of Heads-Up Pennies in Box
0	
1	
2	
3	
4	
5	
6	
7	
8	
9	
10	

Name _____ Class _____ Date _____

GRAPH

Construct a line graph of your data on the grid below. On the horizontal axis, plot the "shake number." On the vertical axis, plot the "number of pennies in box." Connect the points with a line. Give your graph an appropriate title.

Title: _____

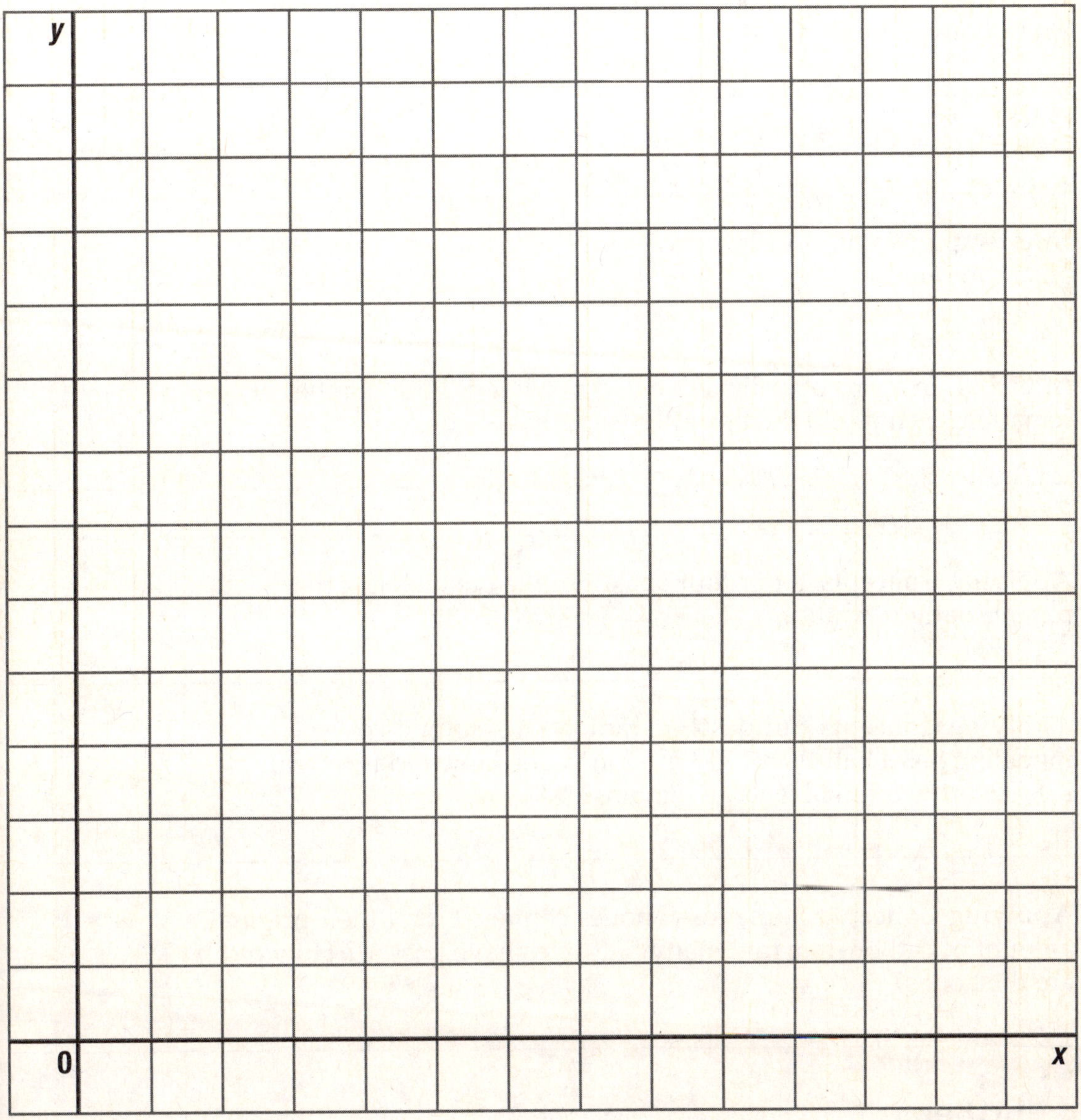

Name _____ Class _____ Date _____

Analysis and Conclusions

1. **Observing** What percentage of the original 100 pennies remained after the first shake of the box? The second shake? The third shake? What fractions do these percentages represent?

   Approximately 50%; 25%; 12.5%; 1/2; 1/4; 1/8

2. **Inferring** How are the above fractions related to the probability of each penny landing heads-up?

   The probability of each penny landing heads up was 1:2 for each shake of the box. Half of the remaining pennies will land tails up with each successive shake. The fractions are halved with each shake.

3. **Inferring** In terms of radioactive decay, what did each shake of the box represent?

   one half-life

4. **Inferring** In terms of radioactive decay, what did the number of remaining pennies in the box after each shake represent?

   The number of remaining pennies represented the radioactive isotope that did not decay—the remaining parent atoms.

6. **Applying Concepts** After four shakes of the box, what is the parent/daughter ratio?

   1:16

7. **Applying Concepts** Suppose the radioactive isotope you are modeling has a half-life of 713 million years. How old is the sample if 1/32 of the original isotope remains?

   3565 million, or 3,565,000,000 years old

8. **Applying Concepts** Some fossil bones contain 1/8 of their original amount of carbon-14. How many half-lives have passed? How old are the bones?

   Three half-lives have passed; the bones are 17,190 (3 × 5730) years old.

Go Further

Use research resources in the library or on the Internet to find at least one instance where geologists or archaeologists used radiometric dating in their work. Write a paragraph describing what they determined, which radioactive isotope was used for the dating procedure, and why it was important to determine the age of the object(s) they measured.

Answers will vary, but students should include in their paragraphs the half-life of the radioactive isotope used in dating.

Earth Science Lab Manual

Name _____ Class _____ Date _____

Chapter 13 Earth's History

Investigation 13

Determining Geologic Ages

Introduction

Evidence of past life on Earth can be found in the fossil record. **Fossils** are among the most important tools scientists use to interpret Earth's history. Not only can they help in dating rock layers, they also reveal the changing nature of life over the vast scale of Earth's history. Fossils and **relative** and **absolute dating** have also told us what we know about geologic changes on Earth—from the gradual rearrangements of the continents to cataclysms that caused mass extinctions.

In this investigation, you will try your hand at using fossils, relative dating, and radiometric dating to uncover some of Earth's history.

This activity encompasses information found in both Chapters 12 and 13. Have students review the sections in Chapter 12 on Principles of Relative Dating, Index Fossils, and Radiometric Dating. The Geologic Time Scale in Chapter 13 will also be an important reference for this activity. It is also reproduced in the DataBank.
SKILLS FOCUS: Observing, Classifying, Using Tables and Graphs
TIME REQUIRED: 45 minutes

Problem

How can you interpret the fossil record to determine Earth's history?

Pre-Lab Discussion

Read the entire investigation. Then work with a partner to answer the following questions.

1. **Inferring** This activity incorporates information from Chapters 12 and 13. How are these two chapters related?

   Chapter 12 covers the physical changes on Earth over time. Chapter 13 offers a view

   of the changing nature of life that accompanied the physical changes.

   _____

2. **Inferring** Why is it important to have more than one dating technique available?

   Not all rock layers can be radiometrically dated. The few that can be dated provide reference points

   for relative dating and index fossil techniques.

   _____

3. **Using Analogies** Look at the Geologic Time Scale (Resource 10) in the DataBank. How are eras, periods, and epochs like the divisions used in textbooks?

   The chapters are like eras, the sections are like periods, and the paragraphs are like epochs.

   _____

Earth Science Lab Manual ▪ 93

Name _____ Class _____ Date _____

4. Posing Questions Write a question that summarizes the purpose of this activity.

Sample question: How are geologic dating techniques combined to sequence and date events in Earth's history?

Materials *(per pair of students)*
geologic block diagram (Figure 1)
logarithmic scale showing decay of U-235
Resource 10 in the DataBank
Resource 11 in the DataBank

Procedure
Part A: Understanding Relative Dating
1. Carefully study Figure 1, the geologic block diagram below. Use the rules you have learned for determining relative age to find the sequence of geologic events. List their letters from oldest to youngest in the space provided beside the figure.

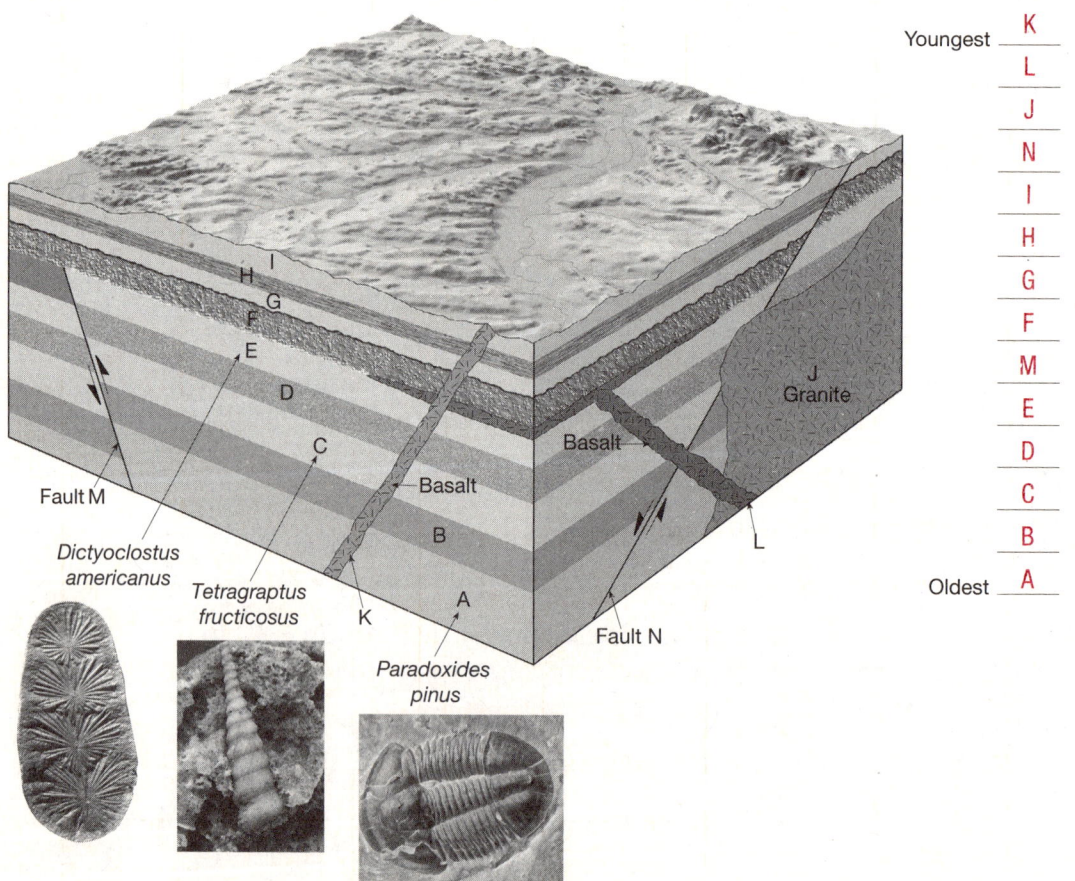

Youngest K
L
J
N
I
H
G
F
M
E
D
C
B
Oldest A

Figure 1

Earth Science Lab Manual ▪ 94

Name _____ Class _____ Date _____

Part B: Understanding Half-Life

2. Study Data Table 1 below. It contains information about the parent-daughter ratios of the isotope uranium-235 (U-235) for several of the rock layers in the block diagram.

DATA TABLE 1

Parent-Daughter Percentages of Isotope U-235			
Rock Layer	Percentage of U-235	Absolute Age	Period
G	94	70 my	Cretaceous
F	90	150 my	Jurassic
D	65	380 my	Devonian
B	60	500 my	Cambrian

3. Study the graph Half-Life of U-235 below. The half-life graph is plotted on a logarithmic scale, which straightens the curved line for radioactive decay. This scale can make it easier to plot data, as well as easier to use when the parent-daughter ratio represents less than a single half-life. Use the graph to determine the absolute ages of the rock layers in the chart.

Half-Life of U-235

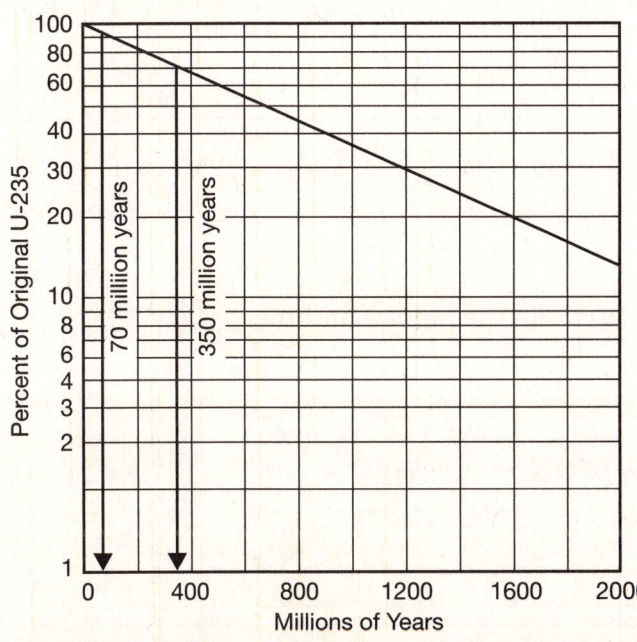

4. It takes 713 million years for half of a sample of U-235 to decay to lead-207. Use the Geologic Time Scale (Resource 10 in the DataBank) to complete Date Table 1 with the period during which each rock layer formed.

Earth Science Lab Manual ▪ 95

Name _____ Class _____ Date _____

Part C: Understanding Index Fossils

5. Complete Data Table 2 below using the Geologic Time Scale (Resource 10) and the Key to Index Fossils (Resource 11) in the DataBank to determine approximate absolute ages for the rock layers in the block diagram (Figure 1) that display index fossils.

DATA TABLE 2

Approximate Age of Index Fossils			
Rock Layer	**Index Fossil**	**Period**	**Approximate Age**
C	Tetragraptus fructicosus	Ordovician	443–490 my
E	Dictyoclostus americanus	Pennsylvanian	290–323 my
A	Paradoxides pinus	Cambrian	490–540 my

Analysis and Conclusions

1. **Applying Concepts** Which law, principle, or doctrine of relative dating did you apply to determine the relative ages of rock layers H and I?

   law of superposition

2. **Applying Concepts** Which law, principle, or doctrine of relative dating did you apply to determine the relative ages of fault M and rock layer F?

   cross-cutting relationships

3. **Applying Concepts** Explain how you know that fault N is older than the igneous intrusion J.

   The intrusion cuts through the fault.

4. **Inferring** Why are there no index fossils in the granite and the basalt?

   Igneous rocks form from solidified lava or magma; therefore, they contain no fossils.

5. **Applying Concepts** How did you determine the sequence of the three igneous intrusions?

   The youngest is K because it cuts across all layers. Layer L is the next youngest. Layer L is older than K because K cuts across L. Layer K is younger than J because L cuts across J.

6. **Problem Solving** How is it possible for two distinct rock layers to derive from the same period?

   Rock layers do not necessarily match time scale categories, which are based on life forms.

Name _____ Class _____ Date _____

Chapter 14 The Ocean Floor   Investigation 14

Modeling the Ocean Floor

Have students review the information in their textbooks on satellite bathymetry. **SKILLS FOCUS:** Observing, Measuring, Classifying, Using Models
TIME REQUIRED: 45 minutes

Introduction

The elevation of mountains is always expressed in terms of sea level. For example, Mount Everest rises 8848 m above sea level. But did you know the ocean floor is not actually level? Ocean basins have a variety of features including chains of volcanoes, tall mountain ranges, trenches, and large submarine plateaus.

Measuring the ocean floor from space has led scientists to a better understanding of the ocean floor. Data from orbiting satellites like the *TOPEX/Poseidon* and *Jason-1* reveal small-scale differences in ocean-surface height caused by ocean-floor features. Gravity attracts water toward regions where massive ocean-floor features occur—mountain ranges produce elevated areas of the ocean surface, and trenches cause slight depressions. Satellites are able to measure these small differences by bouncing microwaves off the ocean surface.

Scientists can use the sea-surface measurements produced by satellites and traditional sonar depth measurements to create detailed maps of ocean-floor features. These maps aren't generally accurate enough to produce navigational charts. But they are useful for many other purposes such as determining how undersea structures affect ocean currents and finding shallow seamounts where fish might be abundant.

In this investigation, you will create a three-dimensional model of the ocean floor, using a map derived from satellite data.

Problem

How can you model the ocean floor, using maps created from satellite data?

Pre-Lab Discussion

Read the entire investigation. Then work with a partner to answer the following questions.

1. **Posing Questions** Write a question that summarizes the purpose of this investigation.

   Sample question: How can a map of the ocean floor derived from satellite data be used to create

   a cardboard contour model of ocean-floor topography?

   _____

2. **Predicting** How will you represent each depth on your model?

   Each depth will have a different color, and it will rise to a different height on the model.

   _____
   _____

Earth Science Lab Manual ▪ 97

Name _____ Class _____ Date _____

3. **Designing Experiments** What is the purpose of the small pieces of corrugated cardboard you will place between the contour layers on your model?

   The corrugated cardboard will help to show relief, or the highs and lows of the ocean floor.

4. **Inferring** What is one advantage of the model? What is one disadvantage of the map?

   One advantage of the model is that it shows the ocean-floor topography in three dimensions.

   One disadvantage of the map is that it only shows the shape of the ocean-bottom topography in

   two dimensions.

5. **Inferring** Based on the data on the map, do you believe you will be able to create an accurate model of the ocean floor? Explain your answer.

   Students should realize that their models will not be entirely accurate because the original map

   is not entirely accurate. Students will be limited in their ability to depict the height of ocean-floor

   features.

Materials *(per pair of students)*

Resource 4 in the DataBank
Resource 5 in the DataBank
tracing paper
thin cardboard
thick, corrugated cardboard
large piece of cardboard (9 in. × 11 in.)
scissors
glue
tempera paint
plastic gloves

Safety

Put on a laboratory apron. Use care with the scissors, especially when cutting through corrugated cardboard. Wear plastic gloves and take care when using the paints. Wash your hands after using the paints. Note all safety symbols next to the steps in the Procedure and review the meanings of each symbol by referring to the symbol guide on page xiii.

Name _____ Class _____ Date _____

Procedure

1. Place the tracing paper over the image of the mid-ocean ridge in the Southern Ocean. Trace around each color. Each tracing represents one contour, which means the depth for that area is the same.
2. Use the scale to determine the depth (in km) represented by each color. Write the depth each of your tracings represents.
3. Trace each contour line from the tracing paper onto a piece of thin cardboard. Cut out each one.
4. Put on your lab apron and rubber gloves. Choose a color to represent each depth. Paint each area the color you have chosen to correspond to that depth.
5. On the side of the large piece of cardboard, make a key for the color you chose to represent each depth. You will build your model on this piece of cardboard. Wash your hands when you have finished painting.
6. Cut out small squares of corrugated cardboard to place between each layer of your model.
7. Create your model by stacking each contour in order of depth. Place corrugated cardboard between the contour layers to provide relief.
8. Carefully glue your model together.

Analysis and Conclusions

1. **Observing** Check the scale of your model. How do the numbers correspond with height of the ocean floor features? Explain your answer.

   The higher the features are, the lower the numbers are. The numbers represent depth below sea level rather than the height above the ocean bottom.

2. **Inferring** Imagine you made a similar model from a topographic map of an area of land. How would these numbers correspond with the height of the land features? Explain your answer.

   For a land map, the higher the features are, the higher the numbers are. The numbers represent the height above sea level rather than the depth below the ocean surface.

3. **Observing** What is the contour interval for your model?

   depth in kilometers; for example, 10 km

Name _____ Class _____ Date _____

4. **Inferring** The map from which you created your model was derived from satellite measurements of sea-surface levels. These levels vary by only about 60 cm. How do you think scientists can determine these depth differences in kilometers?

   Each centimeter difference in sea level represents a much larger difference in ocean depth.

   Scientists have determined a method for converting the sea-level differences to depth differences.

5. **Inferring** Large-scale ocean currents and eddies also affect the height of the sea surface. How do you think scientists can gain reliable information about the ocean floor topography from satellite data?

   Scientists must have the ability to subtract the effects of currents and eddies.

6. **Applying Concepts** How could scientists check to be certain that the ocean-floor maps they are generating from satellite data are accurate?

   Scientists can compare maps generated by satellite data with maps generated by sonar data.

   They can also compare the maps to observations made by submersibles.

7. **Applying Concepts** How could you have improved the model you made in this exercise?

   Answers will vary, but students may say that they could have used different maps of the area or

   additional contours to create more detail.

Go Further

Using the library or Internet, research the value of satellite-generated data for determining the motion of currents in the world oceans. How is the motion of currents related to ocean-floor topography?

Students will find that satellite data provide more complete current information than had been available before. Most currents originally were charted by sailors. Now, with satellite data, scientists can view changes in ocean currents every few days. The data are invaluable for purposes such as predicting El Niño events. Scientists also have learned from the data how ocean-floor features affect currents that otherwise would be affected only by the shape of the ocean basins and the rotation of Earth.

Name _____ Class _____ Date _____

Chapter 16 The Dynamic Ocean   Investigation 16

Shoreline Features

Refer students to the section in their textbooks that discusses shoreline features and the methods humans use for stabilizing the shore. **SKILLS FOCUS:** Observing, Using Models
TIME REQUIRED: 35 minutes

Introduction

Shorelines come in many forms—from the steep Pacific cliffs to sandy barrier beaches along the Atlantic. Geologists have developed two general classifications for coasts that are based on how the land is affected by past changes in sea level.

Emergent coasts result from rising land or falling sea level. They are characterized by **wave-cut cliffs** or **platforms. Submergent coasts** result from rising sea level or sinking land. These often feature **estuaries** resulting from flooded river mouths. Both coast types experience wave erosion and deposition of sediments. Therefore, you can find similar features in all coastal areas. Some of these features include **beaches, spits, tombolos,** and **baymouth bars.**

In this investigation, you will identify coastal features and analyze how they formed. You will also determine how man-made features designed to stabilize the shore affect natural erosion and deposition processes.

Problem

How can you identify and analyze shoreline features?

Pre-Lab Discussion

Read the entire investigation. Then work with a partner to answer the following questions.

1. **Posing Questions** Write a question that summarizes the purpose of this investigation.

   Sample question: How can students become more familiar with shoreline features and the

   processes that form them?

2. **Inferring** How are the diagrams, maps, and aerial views helpful to this investigation?

   The overhead view makes it easy to place the features into the context of overall coastal processes.

3. **Inferring** Why do all shorelines share similarities?

   The same processes of erosion and deposition act upon all coastal areas.

Earth Science Lab Manual ▪ 101

Name _____ Class _____ Date _____

Materials *(per pair of students)*
Figures 1, 2, and 3

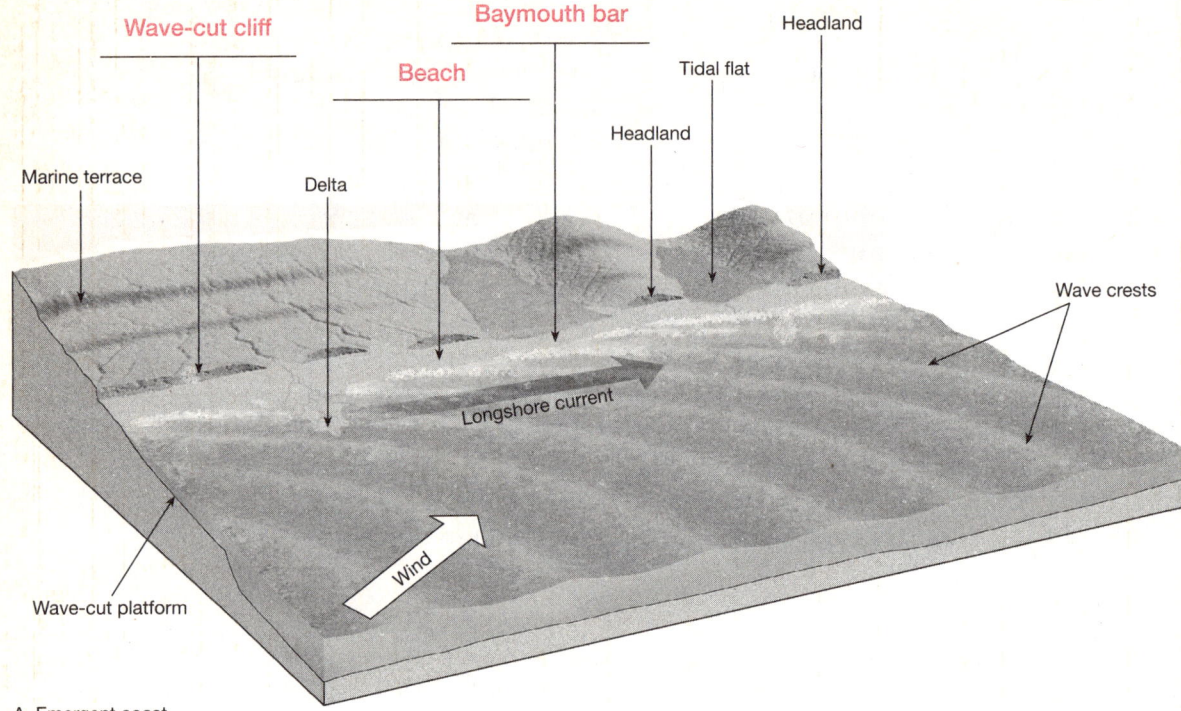

A. Emergent coast

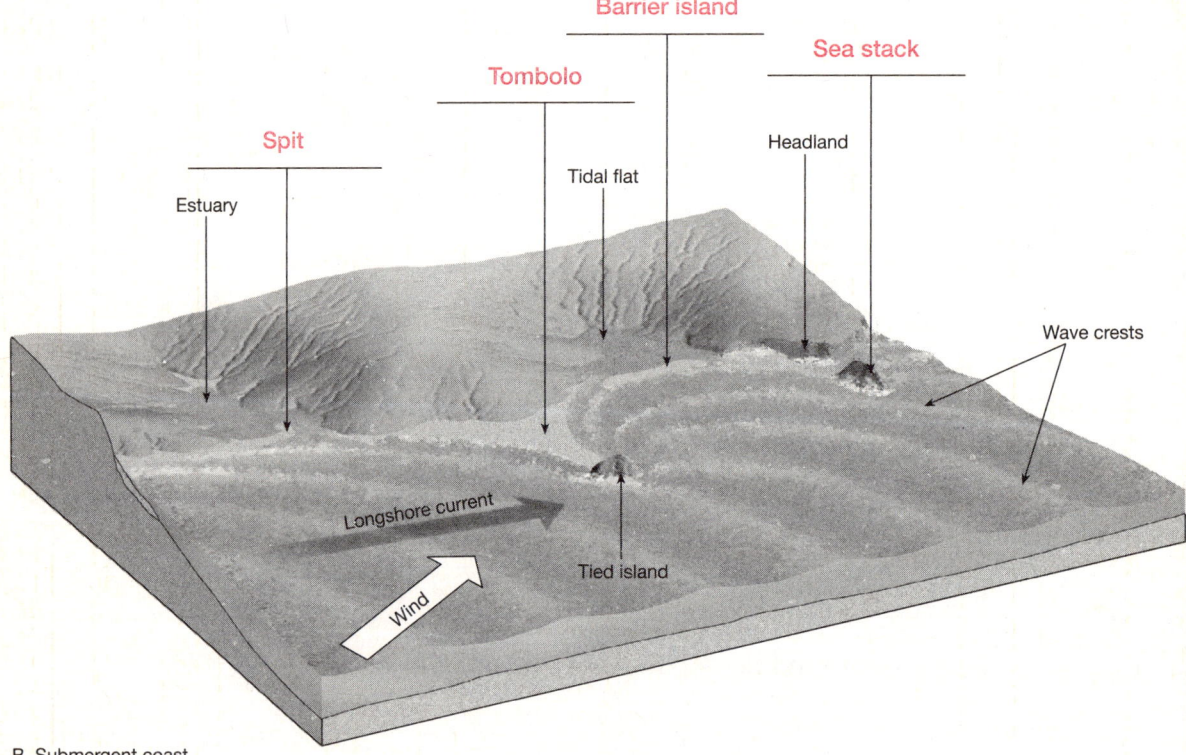

B. Submergent coast

Figure 1

Name _____ Class _____ Date _____

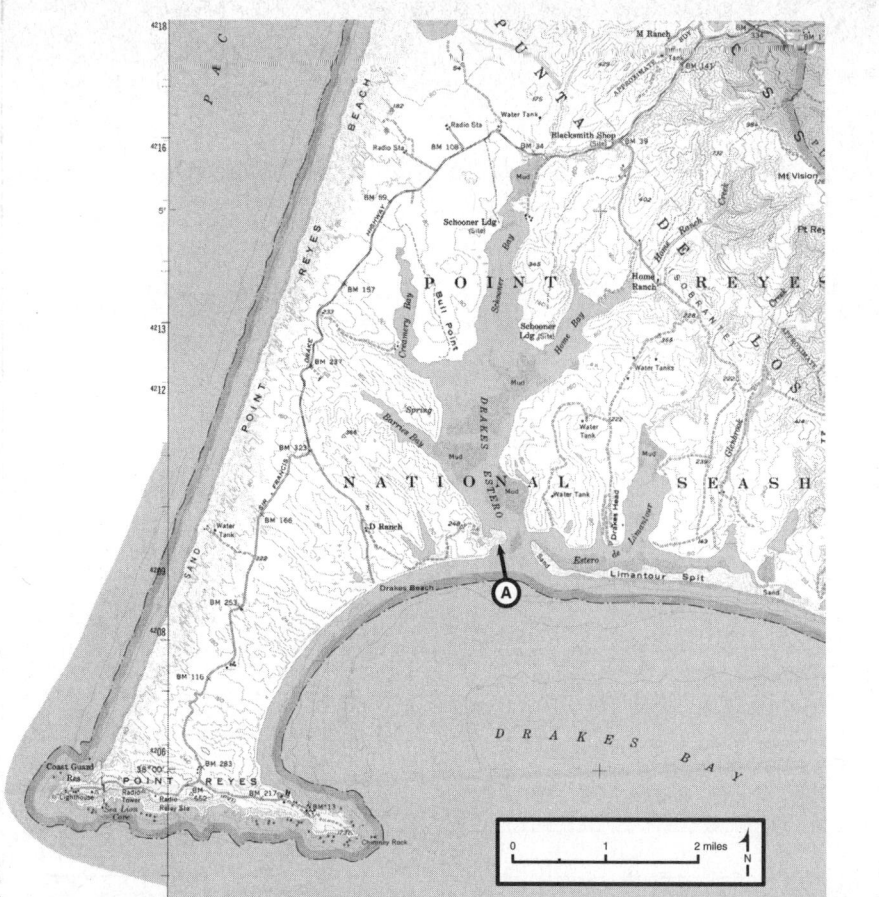

Figure 2

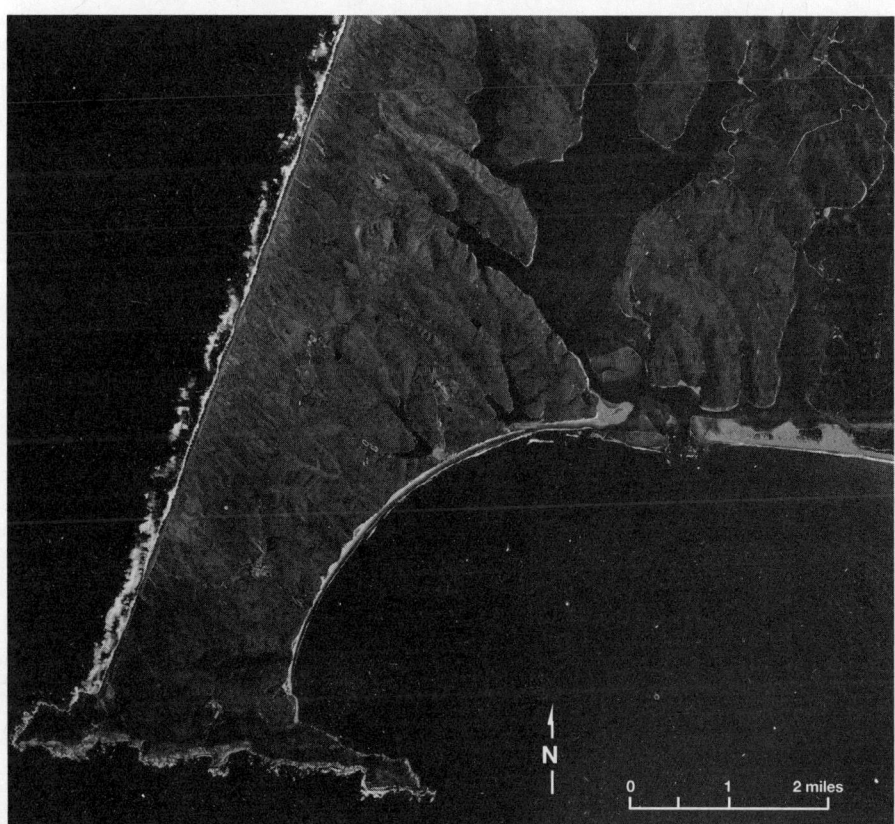

Figure 3

Earth Science Lab Manual ▪ 103

Name _____ Class _____ Date _____

Procedure

Part A: Identify Coastal Features

1. Look at Figure 1.
2. Identify each of the following coastal features by writing the name in the space above the appropriate arrow: *barrier island, tombolo, beach, spit, sea stack, baymouth bar, wave-cut cliff.* See Figure 1 for answers.
3. Study the illustrations in Figure 1. Use them to complete the Data Table to identify which process formed each feature.

DATA TABLE

Classifying Shoreline Features	
Feature	**Process**
sea stack	erosion
wave-cut cliff	erosion
delta	deposition
beach	deposition
wave-cut platform	erosion
marine terrace	erosion
baymouth bar	deposition
headland	erosion
spit	deposition
tombolo	deposition
barrier island	deposition

Name _____ Class _____ Date _____

Part B: Identify Current Directions

4. Contrast the topographic map of a portion of Point Reyes, California, in Figure 2 with the high-altitude image of the same area in Figure 3.

5. Indicate the direction of the current in the vicinity of Limantour Spit by drawing a large arrow.   The current direction is east to west.

Analysis and Conclusions

1. **Observing** Explain how you determined the current direction near Limantour Spit. Use Figures 2 and 3 to explain your answer.

   There is very little sand inside the point itself. The hook at the end of the spit also indicates that the sand is moving from east to west.

2. **Observing** Is the Point Reyes area a submergent or emergent coast? Explain your reasoning using specific features on the map and image.

   The extensive estuary called Drake's Estero is a clear indication that this area is a submergent coast. Its irregular appearance also is indicative of flooding that occurs in a submergent area. The area also lacks terracing, which you would expect to see on the shore if the area were emergent.

3. **Applying Concepts** Imagine you were an engineer seeking to restore boat access to the lagoon in the emergent coast in Figure 1. What structure would you build and where would you build it? Explain your answer.

   A groin built on the beach on the left side of the lagoon would prevent sand from migrating with the longshore current across the entrance to the lagoon. Wave erosion would eventually open the beach and allow access to the lagoon, which would then be called an estuary.

4. **Applying Concepts** Point Reyes, a typical headland, is undergoing severe wave erosion. What type of features are Chimney Rock and the other rocks located off the shore of Point Reyes? How did the features form?

   The rocks are sea stacks. They formed after wave erosion on the opposite sides of the headland formed a sea arch, which collapsed and left an isolated remnant of the headland.

Name _____ Class _____ Date _____

5. Predicting What will eventually happen to the land at the tip of Point Reyes?

Severe erosion from the refracted waves of the Pacific Ocean will continue. Further sea caves will form in the point. If they unite, they will form arches. More erosion will cause the arches to fall in and form sea stacks.

6. Inferring Imagine somebody constructed a groin by the word "Limantour" on Limantour Spit. On which side of the groin, east or west, will sand accumulate? What will be the effect on the opposite side of the groin?

Sand will accumulate on the east side of the groin. The west side will be "sand starved" and eroded.

7. Inferring How did the U-shaped lake east of D Ranch form?

The lake is an estuary that has been closed off by the formation of a baymouth bar at its mouth.

8. Applying Concepts What is the feature labeled *A* on the map?

a spit

Go Further

Coastal features are either helpful or harmful to navigation. Use the library or the Internet to find either an instance of a boater who ran into trouble due to a coastal feature or a boater who was saved by the presence of a coastal feature. Report your findings in the form of a news article.

Students will find instances of surf building up at baymouth bars lying close beneath the surface. These features, such as one at the mouth of the Columbia River, can be deadly to mariners. On the other hand, the lagoons and estuaries behind bars can save boaters from the disastrous effects of hurricanes. Tidal flats are sometimes utilized by sailors who want to clean the hulls of their boats. They sail up during high tide, and they are left high and dry to do their work at low tide.

Name _____ Class _____ Date _____

Chapter 17   The Atmosphere: Structure and Temperature   Investigation 17A

Determining How Temperature Changes with Altitude

Have students review the information on the composition and structure of the atmosphere in their textbooks.
SKILLS FOCUS: Using Graphs, Inferring, Calculating   **TIME REQUIRED:** 35 minutes

Introduction

The atmosphere is divided into four layers based on temperature: the **troposphere,** the **stratosphere,** the **mesosphere,** and the **thermosphere.** The temperature in the lower 12 km of the atmosphere decreases with altitude. However, at altitudes from about 12 to 45 km, the temperature increases.

In this investigation, you will explore the temperature changes in Earth's atmosphere as altitude increases and investigate what causes these temperature changes.

Problem

How does the temperature of Earth's atmosphere change with altitude?

Pre-Lab Discussion

Read the entire investigation. Then work with a partner to answer the following questions.

1. **Posing Questions** Write a question that summarizes the purpose of this investigation.

   Sample question: How does the temperature of Earth's atmosphere change with altitude?

   _____

   _____

2. **Inferring** What are the possible sources of heat for the atmosphere?

   The possible sources of heat for the atmosphere include conduction, radiation, and convection from

   Earth's surface to the atmosphere. The absorption of solar radiation from the sun by the atmosphere

   also is a source of heat.

3. **Predicting** What substance in the upper atmosphere is important to temperature changes in the upper atmospheric layers?

   Answers will vary, but most students should predict that oxygen and nitrogen are important in

   causing temperature changes in the upper atmosphere. They absorb short-wave, high-energy

   solar radiation.

Earth Science Lab Manual ▪ 107

Name _____ Class _____ Date _____

Materials *(per group of students)*
ruler or straight edge
colored pencils
tracing paper
Resource 12 in the DataBank

Students can use the completed figure as a study aid.

Procedure
1. Carefully study the Atmospheric Temperature Curve shown in Resource 12.
2. Using tracing paper and the ruler, trace Resource 12.
3. Use the ruler to draw in the lines for the tropopause, stratopause, and mesopause. Label each line. If necessary, use your textbook as a reference.
4. Label the troposphere, mesosphere, stratosphere, and thermosphere.
5. Shade in each section. Use a different color for each section.

Analysis and Conclusions
1. **Using Graphs** What is the approximate temperature of the atmosphere at each of the following altitudes?

   10 km: _____−58_____ °C
   50 km: _____−2_____ °C
   80 km: _____−79_____ °C

2. **Using Graphs** How does the temperature change with altitude in the troposphere?

   The temperature decreases with increasing altitude in the troposphere.

3. **Drawing Conclusions** What causes the temperature change in the troposphere?

   The source of the heat in the troposphere is Earth's surface. Most of the heating occurs at the bottom of the atmosphere and it decreases with increasing altitude.

4. **Using Graphs** How does the temperature change with altitude in the stratosphere?

   The temperature increases in the stratosphere as altitude increases.

Earth Science Lab Manual ▪ 108

Name _____ Class _____ Date _____

5. Drawing Conclusions What causes the temperature change in the stratosphere?

The ozone in the stratosphere absorbs ultraviolet radiation from the sun. This absorption causes the stratosphere to heat up and the temperature to increase.

6. Using Graphs How does the temperature change with altitude in the mesosphere and thermosphere?

In the mesosphere, the temperature decreases with increasing altitude. In the thermosphere, the temperature increases with increasing altitude.

7. Drawing Conclusions Explain the temperature change with altitude in the thermosphere.

The increase in temperature is caused by the absorption of very short-wave, high-energy solar radiation by the atoms of oxygen and nitrogen in the thermosphere.

8. Calculating If the average normal temperature decrease with altitude in the troposphere is 6.5°C/km, calculate the approximate temperature at 6,000 m if the surface temperature is 16°C. Show your work.

> 6,000 meters = 6 kilometers; multiply the rate of temperature decrease 6.5°C/km by 6 km to get a temperature change of 39°C; then subtract 39°C from the temperature at the surface 16°C, (16°C) − (−39°C) = −23°C

9. Calculating If the average or normal temperature decrease with altitude in the troposphere is 6.5°C/km, calculate the approximate altitude in which a pilot would expect to find each of the following atmospheric temperatures, if the surface temperature is 27°C. Show your work.

10°C: _____2600_____ meters

0°C: _____4200_____ meters

> A final temperature of 10°C represents a change of 17°C. 27°C − 10°C = 17°C; then divide the temperature change by the rate of temperature change, 6.5°C/km; 17°C ÷ 6.5°C/km = 2.6 km or 2600 meters, so it is 10°C at approximately 2600 meters above the surface. A final temperature of 0°C represents a change of 27°C. 27°C − 0°C = 27°C; then divide the temperature change by the rate of temperature change; 27°C ÷ 6.5°C/km = 4.2 km − 4200 meters; so it is 0°C at approximately 4200 meters above the surface.

10. **Inferring** Of what importance is the gas ozone in the stratosphere? How would a decrease of ozone in the stratosphere affect the radiation received at Earth's surface?

Ozone in the stratosphere absorbs ultraviolet radiation from the sun. A decrease of ozone in the stratosphere would result in more ultraviolet radiation from the sun reaching Earth's surface.

Go Further

Temperature measurements from the upper atmosphere are gathered by using weather balloons. The balloons collect data on temperature, humidity, and wind. The temperature data are plotted versus pressure/height on plots called Skew-T diagrams that provide information on the vertical structure of the atmosphere. Compare diagrams from different areas to help answer the following questions. Is the vertical temperature profile of the atmosphere the same everywhere at all times? What can cause the temperature profile to change?

The data and information to investigate the changes in temperature versus altitude in different locations can be found on

For: Chapter 17 Resources
Visit: PHSchool.com
Web Code: cjk-9999

No, the vertical temperature profile of the atmosphere is not the same everywhere at all times. The temperature profile can change due to changes in the air pressure, humidity, and temperature near the surface as the result of the movement of storms or weather fronts.

Name _____ Class _____ Date _____

Chapter 17   The Atmosphere: Structure and Temperature   Investigation 17B

Investigating Factors That Control Temperature

Have students review the information on the heating of the atmosphere and the factors that influence temperature in their textbooks.
SKILLS FOCUS: Using Maps, Inferring, Analyzing Data, Graphing **TIME REQUIRED:** 35 minutes

Introduction

One summer day, the official temperature in Columbus, Ohio, was reported as 88°F. However, the electronic sign at a local drugstore reported a temperature of 97°F. Was the temperature on the sign wrong? Actually, both measurements were correct, but they were measured under different conditions.

Official temperatures are measured in the shade, over a grassy surface, and five feet above the ground. The store's temperature was measured by a sensor in full sun, located close to a dark, paved surface. This difference in measuring the temperature accounted for almost a 10-degree increase in temperature—on the same day and at the same time.

In this investigation, you will explore the differences in temperature across North America and investigate the factors that influence temperature.

The data and information you will need for the investigation can be found on

Go Online PHSchool.com

For: Chapter 17 Resources
Visit: PHSchool.com
Web Code: cjk-9999

Students can complete the lab using the temperature contour plots and temperature change and heat index plots on Resources 14 and 15 in the DataBank. The lab can also be completed using current data, which can be accessed through the Web site listed above. If current data are used, some of the specific questions and location will need to be modified to fit the data plots used. Refer to the new National Weather Service wind chill chart, which was revised in 2001.

Problem

How does temperature vary and what causes these variations?

Pre-Lab Discussion

Read the entire investigation. Then work with a partner to answer the following questions.

1. **Posing Questions** Write a question that summarizes the purpose of this investigation.

   Sample question: What factors influence the differences in temperature across North America?
   _____
   _____

2. **Inferring** What factors can influence temperature?

   Answers will vary, but they could include latitude, location of temperature measurements, altitude,
   geographic position, cloud cover, ocean and wind currents, and type of the land surface.
   _____

3. **Predicting** Which heats up faster, land or water?

   Answers will vary, but most students should predict that land heats up faster and to a higher
   temperature than water.
   _____

Earth Science Lab Manual ▪ 111

Name _____ Class _____ Date _____

Materials *(per group of students)*
ruler or straight edge
graph paper
Resources 14 and 15 in the DataBank

Student may use current data, which can be accessed through the Web site listed on the first page, to complete the lab.

Procedure

1. Study Resource 14. The top map shows surface temperature across the United States, and it has been contoured to show areas that have the same temperature range. The contour lines are similar to the contour lines shown on a topographic map. However, these contour lines represent temperature rather than elevation.

2. Locate your state on Resource 14. Use the scale below the map to determine the temperature range or ranges that occurred in your state on August 9, 2004.

   Answers will vary according to the state in which your school is located.

3. Make a temperature profile, or cross section, from the southwest corner of New Mexico through the Four Corners to the northeast corner of Colorado. The Four Corners is the location where the borders of New Mexico, Arizona, Utah, and Colorado intersect. First place a piece of paper on the United States map on Resource 14, extending from the SW corner of New Mexico to the NE corner of Colorado. Mark the ends of the cross section.

4. Then make marks on the paper where contour lines cross the profile line. Label the areas in between the marks with the corresponding temperature ranges from the map scale.

5. Use the marks to construct a temperature profile across this area on graph paper. Make the horizontal scale of the graph double the horizontal scale of the map.

6. Use the contour plots on Resources 14 and 15 in the DataBank (United States Surface Temperature, North American Surface Temperature, 24-Hour Temperature Change, and Surface Heat Index) to answer the questions in **Analysis and Conclusions.**

Analysis and Conclusions

1. **Using Graphs** How does the temperature profile you made across New Mexico and Colorado change?

   The temperature profile begins at 95°F–100°F at the New Mexico border, decreases to 85°F–90°F in the Four Corners, and continues to decrease to 75°F–80°F at the northeastern corner of Colorado.

Name _____ Class _____ Date _____

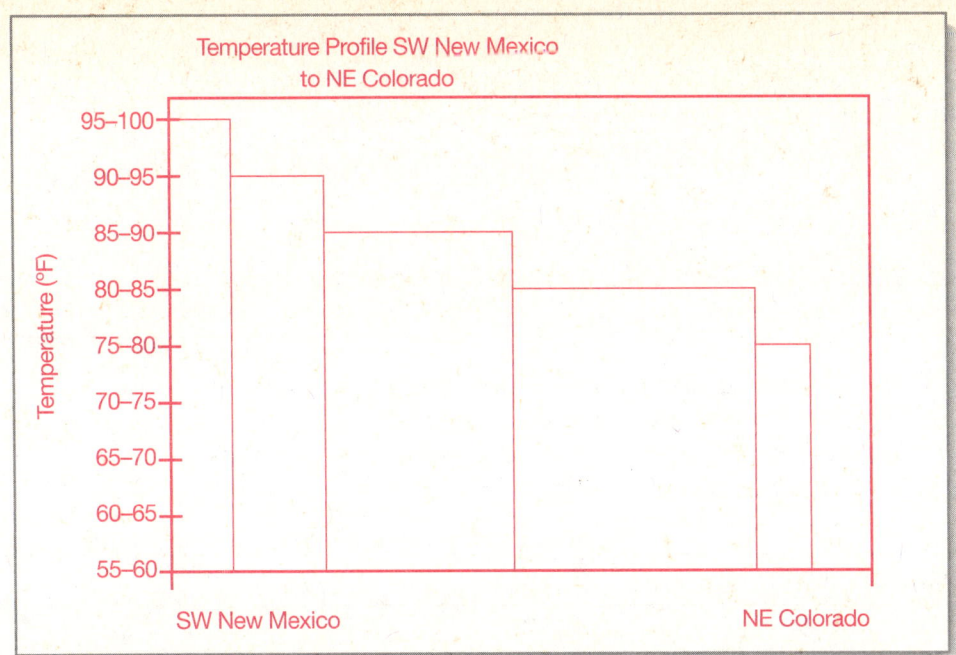

2. Observing What are the lowest temperatures shown on the North American Surface Temperature Contour Plot on Resource 14? What are the highest temperatures?

The lowest temperatures are 35°F–40°F, and the highest temperatures are 110°F–115°F.

3. Inferring Which coast of North America is the leeward coast and which is the windward coast?

The west coast of North America is the windward coast and the east coast is the leeward coast.

4. Inferring Which coast, leeward or windward, usually has cooler temperatures? Explain. Does the data shown on both maps on Resource 14 support this inference?

Windward coasts usually have cooler temperatures because the winds blowing from the ocean moderate the temperatures. Yes, the maps both support this inference. The west coast generally has lower temperatures than the east coast.

5. Analyzing Data What is the general trend of temperature shown on the North American Surface Temperature Contour Plot?

The temperatures are lower in the north and increase toward the south.

Name _____ Class _____ Date _____

6. Analyzing Data On the maps on Resources 14, there is a band of cooler temperatures that extends from northern New Mexico through Colorado, and up into Wyoming, Montana, and Idaho. What do you think could be influencing this area of cooler temperatures?

Answers will vary, but most students should respond that the higher elevations or altitude of the Rocky Mountains are causing the cooler temperatures.

7. Analyzing Data Use the 24-hour temperature change contour plot on Resource 15 to determine where in the United States the temperature change over the 24-hour period was the greatest. Was the temperature change positive or negative?

The greatest temperature change occurred in Oklahoma. The temperature change was positive.

The temperature changed 8 degrees during the 24-hour period.

8. Analyzing Concepts Study the heat index contour plot on Resource 15. The heat index is used to warn people when temperatures are high enough to pose a health hazard. The heat index combines the air temperature with relative humidity to determine the apparent temperature—what the air temperature "feels like" to the average person. A heat index of 90°F–105°F with prolonged exposure or physical activity can cause sunstroke or heat exhaustion, which can be dangerous to health. Where in the United States should people be warned about the possible danger of prolonged outdoor physical activity?

In the southwestern United States (southern California, Nevada, and Arizona), in Louisiana, and parts of Oklahoma, Texas, Arkansas, Mississippi, Alabama, and Florida the heat index is above 90°F.

Go Further

Investigate the current temperatures, heat index, or wind chill in your area by accessing temperature plots available on the Internet. Your teacher will provide you with the Web site information or with copies of the current data. Which features in your area have a strong influence on local temperature?

Answers will vary, depending on the area and time of year, but students might include altitude, latitude, or nearby features, such as mountains or large bodies of water. Advanced students can access historical climate data, such as temperature, precipitation, droughts, etc., to conduct more in-depth investigations. The data and information students need can be found on

For: Chapter 17 Resources
Visit: PHSchool.com
Web Code: cjk–9999

Name _____ Class _____ Date _____

Chapter 18 Moisture, Clouds, and Precipitation Investigation 18

Recipe for a Cloud

Have students review the information on cloud formation in their textbooks. **SKILLS FOCUS:** Observing, Using Models, Calculating, Using Tables and Graphs **TIME REQUIRED:** 45 minutes

Introduction

Clouds are a form of condensation, and they are best described as visible mixtures of tiny deposits of water or tiny crystals of ice. **Condensation** is the process of water vapor changing to a liquid state. For example, on a hot, muggy day, water droplets form on a cool glass of lemonade— that's condensation. **Evaporation** is the process of a liquid changing to a gas at the liquid's surface. A puddle you see in the morning that is gone by the afternoon is an example of evaporation.

When you use a hand pump to inflate a bicycle tire, the metal pump cylinder feels warm when you have finished. Its temperature increases when air is compressed. If you hold your hand just above the tire valve while you release the air, the escaping air feels cool. That cooling is caused by expanding air. This cooling and warming of air is caused by pressure changes called **adiabatic temperature changes.**

A raging forest fire sends plumes of smoke high into the sky. The smoke particles become surfaces to which liquid water can stick. The surfaces are called **condensation nuclei.**

In this investigation, you will examine the formation of clouds and the roles played by temperature changes, humidity, and the presence of tiny particles.

Problem

What are the necessary processes and conditions for cloud formation?

Pre-Lab Discussion

Read the entire investigation. Then work with a partner to answer the following questions.

1. **Forming Hypotheses** Write a hypothesis that explains how pressure, humidity, and the presence of tiny particles in the air contribute to cloud formation.

   For clouds to form, there must be enough water vapor in the air (high humidity), low enough

   temperatures (reduced air pressure), and some particles upon which condensation can occur.

2. **Designing Experiments** Why is it important to use a container in which the pressure can be easily and significantly changed?

   The pressure must be changed significantly so that the water vapor in the air expands and cools

   enough. The expanding and cooling air allows the vapor to condense on any small particles that

   are in the air.

Earth Science Lab Manual ▪ 115

Name _____ Class _____ Date _____

3. Controlling Variables When the cold water is used, what is the independent variable?

Air pressure is the independent variable.

4. Controlling Variables When the hot water is used, how does it affect cloud formation?

The hot water creates more humidity than the cold water. This difference in humidity serves as the independent variable between the cold-water and hot-water parts of the investigation.

Materials *(per group)*
graduated cylinder
gallon glass pickle jar
plastic freezer storage bag (26 cm × 26 cm)
rubber band (15-cm circumference; 0.5-cm width)
cold tap water
hot tap water
safety matches

Begin collecting glass pickle jars in advance of this investigation. You might offer some reward for students who bring them in. Another resource might be the school cafeteria. An alternative chamber for cloud formation is a 2-L plastic soda bottle. Hold the smoldering match beneath the opening of an inverted bottle to introduce particles. Instead of using a plastic bag to change the air pressure, students can simply squeeze the bottle to increase the pressure and then release the bottle to create a sudden pressure decrease. Try this before you assign the investigation so you can offer advice on technique.

Safety
Put on safety goggles. Be careful to avoid breakage when handling glassware. Be careful when using matches. Do not reach over an open flame. Tie back loose hair and clothing when working with flames. Note all safety symbols next to the steps in the Procedure and review the meaning of each symbol by referring to the symbols guide on page xiii.

Procedure

1. Use the graduated cylinder to measure 40 mL of cold water. Pour the water into the jar. Place the plastic bag into the jar so that the top edges of the bag lie just outside the rim of the jar, as shown in Figure 1. **CAUTION:** *Wipe up any spilled liquids immediately to avoid slips and falls.*

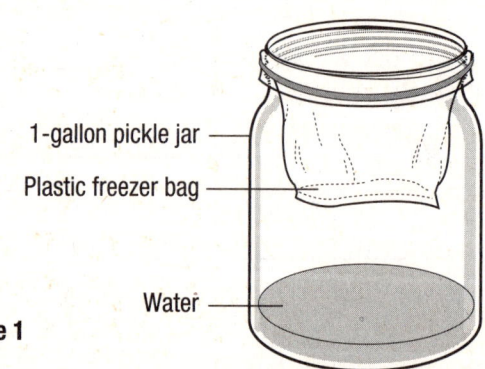

Figure 1

Earth Science Lab Manual • 116

Name _____ Class _____ Date _____

2. Secure the top of the bag to the outer rim of the jar using the rubber band, as shown in Figure 2.

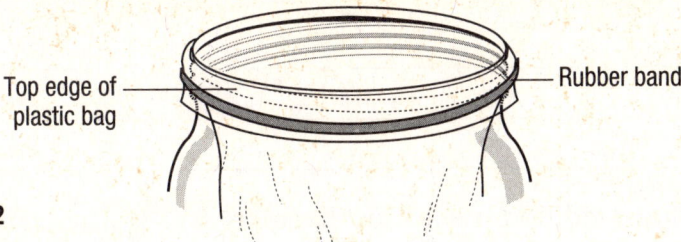

Figure 2

3. Put your hand inside the bag and grab the bottom edge of the bag. Rapidly pull the bag out of the jar. Carefully watch the inside of the jar after pulling the bag out. Record your observations in the Data Table.

4. Remove the rubber band and plastic bag. Light a match and allow it to burn for about 3 seconds. Drop the match into the jar. Quickly place the plastic bag inside the jar and secure its top edge around the rim of the jar with the rubber band.
   CAUTION: *Be careful when using matches.*

5. Repeat Step 3. Carefully watch the inside of the jar after you rapidly pull the bag out of the jar. Record your observations in the Data Table.

6. Remove the plastic bag and rinse out the bottle. Throw away the burned match as directed by your teacher.

7. Repeat Steps 1 through 6 using 40 mL of hot tap water.
   CAUTION: *Be careful not to burn yourself when using hot water and matches.*

Observations

DATA TABLE

Water Type	Smoke	Cloud Formation
Cold	Absent	No
Cold	Present	Yes
Hot	Absent	No
Hot	Present	Yes

If clouds formed under more than one set of conditions, did you observe any difference between the clouds?
Answers will depend on whether or not clouds formed in the absence of smoke. The most common result is that, in the presence of smoke particles, the hot water will produce a somewhat denser cloud than the cold water.

Name _____ Class _____ Date _____

Analysis and Conclusions

1. **Inferring** What effect did pulling the plastic bag out of the jar have on the water vapor inside the jar?

   It reduced the air pressure so that the air and the water vapor rapidly expanded. As they expanded, the temperature in the jar decreased temporarily, and the water vapor was able to condense on the particles in the air when particles of smoke were added to the container.

2. **Drawing Conclusions** What conditions were most likely to produce clouds?

   Cloud formation was slightly better for hot water than for cold water, but both provided sufficient humidity for clouds to form in the presence of smoke. The presence of smoke is important.

3. **Drawing Conclusions** Did your results support your hypothesis? Explain your answer.

   If students predicted that smoke particles were necessary for clouds to form, then their hypotheses were probably supported by their observations.

4. **Using Models** In this investigation, you observed several factors that contributed to cloud formation. How does each of these factors occur in nature?

   When air rises and expands, it experiences the decrease in pressure that causes the temperature to decrease. Evaporation from surface water creates the necessary humidity in the air. Particles of smoke, dust, and pollen occur in the atmosphere to provide the surfaces on which water vapor can condense.

5. **Designing Experiments** How could you improve upon the design of this investigation to better model the process of cloud formation in nature?

   Answers will vary, but students might include a method of evaporating the water after it is introduced into the chamber but before the pressure is changed. They might suggest using a different particle type. Some students might suggest a method of measuring the temperature before and after the pressure change.

Name _____ Class _____ Date _____

Chapter 19 Air Pressure and Wind

Analyzing Pressure Systems

Investigation 19

Have students review the information on pressure centers and winds in their textbooks.
SKILLS FOCUS: Interpreting Diagrams, Analyzing Data, Inferring
TIME REQUIRED: 45 minutes to locate areas of high and low pressure; 30 minutes to analyze data

Introduction
Wind is caused by differences in pressure. Wind forms as air moves from areas of high pressure to areas of low pressure.

In a **low-pressure system** in the Northern Hemisphere, winds blow inward in a counterclockwise direction. Air pressure is lowest in the center of the system.

In a **high-pressure system** in the Northern Hemisphere, winds blow outward in a clockwise direction. Pressure is highest in the center of the system.

In this investigation, you will identify wind patterns and predict the movement of pressure systems.

Problem
How do wind patterns relate to pressure systems?

Pre-Lab Discussion
Read the entire investigation. Then work with a partner to answer the following questions.

1. **Posing Questions** Write a question that summarizes the purpose of this investigation.

   Sample question: How can you predict the movement of pressure systems by studying wind direction?

2. **Designing Experiments** In which direction will your arrows point around low-pressure systems? In which direction will they point around high-pressure systems?

   The arrows will point in a counterclockwise direction around low-pressure systems. They will point in a clockwise direction around high-pressure systems.

3. **Inferring** Why are low-pressure systems associated with areas of air movement?

   Air moves from areas of high pressure to areas of low pressure.

Earth Science Lab Manual ▪ 119

Name _____ Class _____ Date _____

Materials *(per student pair)*
colored pencils

Discuss the Coriolis effect, which is caused by Earth's rotation. Remind students that the Coriolis effect deflects free-moving objects and fluids, such as wind, to the right of their path of motion in the Northern Hemisphere and to the left of their path of motion in the Southern Hemisphere.

Procedure

1. Sketch the outline of your school building in the blank box below.
2. On a windy day, go outside with your partner and stand with the wind to your back. Slowly walk around the school until you reach an area where you feel the air movement increase or a slight counterclockwise swirl. This location is a small area of low pressure.
3. On your drawing of your school, use a red pencil to write an *L* to mark the area of low pressure.
4. To find small areas of high pressure, slowly walk around the school with the wind in your face until you feel the air grow suddenly still.
5. On your drawing of your school, use a blue pencil to write an *H* to mark the area of high pressure.
6. Walk around the entire perimeter of your school, marking areas of highs and lows on your drawing. Stay within 10 meters of the school.
7. On your sketch of the school, draw arrows indicating the movement of air around the low-pressure systems and high-pressure systems.
8. Figure 1 shows wind direction for a large low-pressure system. When the wind is blowing from the north, the center of the low-pressure system lies to the east of you.

Earth Science Lab Manual ▪ 120

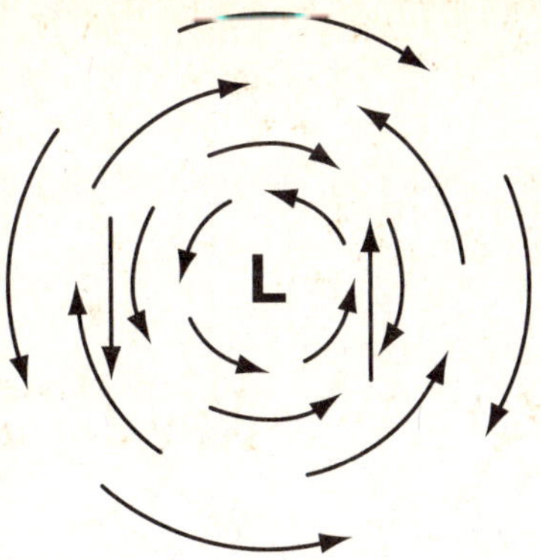

Name _____ Class _____ Date _____

Figure 1

Analysis and Conclusions

1. **Analyzing Data** Describe the general location of low- and high-pressure systems around your school building.

   Answers will vary.

2. **Comparing and Contrasting** Compare your sketches to those of other students. Were the low- and high-pressure systems located in the same places around the school? Why or why not?

   Sample answer: The lows and highs were not in the same places, which probably means that the small pressure systems are not very permanent; they constantly shift.

3. **Interpreting Diagrams** Using Figure 1, if the wind is blowing from the south, where is the low-pressure system relative to your location?

   to the west

4. **Interpreting Diagrams** Using Figure 1, if the wind is blowing from the southwest, where is the low-pressure system relative to your location?

   to the northeast

Name _____ Class _____ Date _____

5. **Inferring** In general, weather systems move from west to east across the United States. Knowing this, which wind direction would indicate that a low-pressure system was headed toward your area?

   A southerly wind direction would indicate that the system was to the west of you and heading east.

Go Further

Obtain a U.S. weather map and a weather forecast for your region from a newspaper or the Internet. Relate the forecast to the presence of high- or low-pressure systems.

Answers will vary. In general, a high-pressure system is associated with fair weather, and a low-pressure system is associated with storms or precipitation.

Name _____ Class _____ Date _____

Chapter 20 Weather Patterns and Severe Storms

Investigation 20A

Analyzing Severe Weather Data

Have students review the information on thunderstorms and tornadoes in their textbooks.
SKILLS FOCUS: Analyzing Data, Interpreting Diagrams, Predicting
TIME REQUIRED: 45 minutes

Introduction

Tornadoes are violent windstorms associated with severe thunderstorms. Meteorologists carefully monitor atmospheric data to predict where thunderstorms might develop. They also attempt to predict whether these storms might spawn powerful tornadoes. To aid them in this task, meteorologists use thermodynamic indices.

Thermodynamic indices are sets of numbers that indicate the state of the atmosphere at a given time and place. Three important thermodynamic indices are the dew-point index, the lifted index, and the storm relative helicity index.

The **dew-point index** indicates the amount of moisture in the atmosphere. The dew-point temperature of an area usually needs to be at least 50°F for a tornado to develop.

The **lifted index** indicates how fast or slow air is rising or sinking. Air must be rising for a thunderstorm—and therefore, a tornado—to develop.

The **storm relative helicity index** indicates whether or not the air is rotating. For a tornado to develop, air must be turned or spun as it rises.

In this investigation, you will use thermodynamic indices and weather maps to predict where a tornado might strike.

Problem

Where are tornadoes most likely to occur?

Pre-Lab Discussion

Read the entire investigation. Then work with a partner to answer the following questions.

1. **Posing Questions** Write a question that summarizes the purpose of this lab.

   Sample question: How can you use thermodynamic indices to predict the formation of tornadoes?

2. **Forming Definitions** What is the dew-point index? For the purposes of this investigation, how does it relate to tornadoes?

   The dew-point index indicates how much moisture is in the atmosphere. The dew-point temperature

   usually needs to be at least 50°F for a tornado to develop.

Name _____ Class _____ Date _____

3. **Formulating Hypotheses** What conditions are most favorable for the development of a tornado?

   <u>The dew-point temperature should be at least 50°F, and the air must be rotating as it rises.</u>

Materials (per pair of students)
3 colored pencils

Procedure

Part A: Analyzing Dew-Point Index
1. Study the weather map in Figure 1. The map shows dew-point temperatures in the United States on April 6, 2003.
2. Choose a colored pencil and shade in the states that have dew-point temperatures that are conducive to the formation of tornadoes.
3. Make a list of the shaded states. If you need help with the names of the states, use Figure 4, the labeled map of the United States.

   <u>Texas, Louisiana, Arkansas, Mississippi, Alabama, Georgia, Florida, South Carolina</u>

Part B: Analyzing Lifted Index
4. Study the weather map in Figure 2. The map shows the lifted index for the United States on April 6, 2003. Data Table 1 includes a scale for the lifted index.
5. Choose a different colored pencil and shade in the states whose lifted indices are conducive to the formation of tornadoes.
6. Make a list of the shaded states. If you need help with the names of the states, use Figure 4, the labeled map of the United States.

   <u>Texas, Louisiana, Mississippi, Alabama</u>

Part C: Analyzing Storm Relative Helicity Index
7. Study the weather map in Figure 3. The map shows the storm relative helicity index for the United States on April 6, 2003. Data Table 2 includes a scale for this index.
8. Choose a different colored pencil and shade in the states whose storm relative helicity values are conducive to the formation of tornadoes.
9. Make a list of the shaded states. If you need help with the names of the states, use Figure 4, the labeled map of the United States.

   <u>Texas, Oklahoma, Louisiana, Arkansas, Missouri, Illinois, Alabama, Mississippi</u>

Earth Science Lab Manual

Name _____ Class _____ Date _____

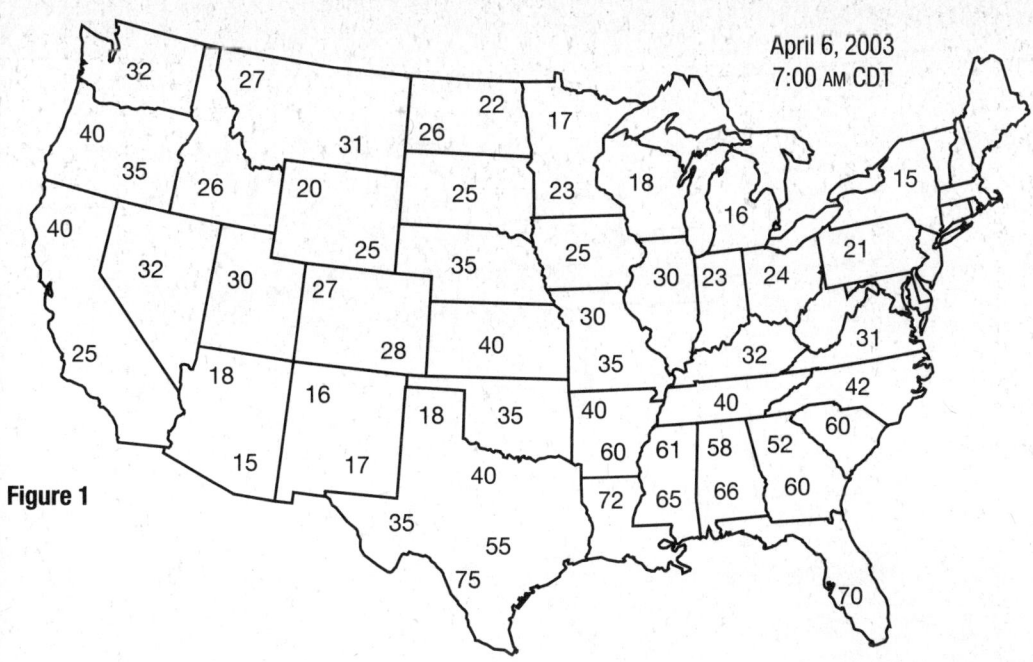

Figure 1

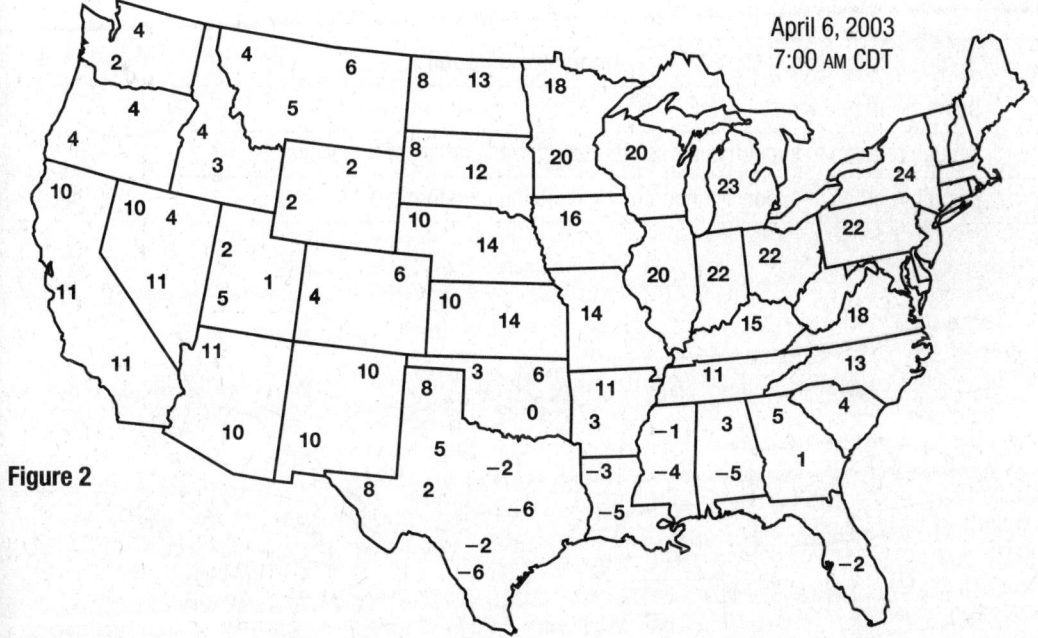

Figure 2

DATA TABLE 1

Lifted Index	Stability
> 0	Air is sinking; very stable atmosphere
0	Stable atmosphere
−1 to −3	Slightly unstable (severe thunderstorms most likely)
−4 to −5	Unstable (severe storms, hail, maybe smaller tornadoes)
< −6	Very unstable (severe storms, larger hail, possibly larger tornadoes)

Name _____ Class _____ Date _____

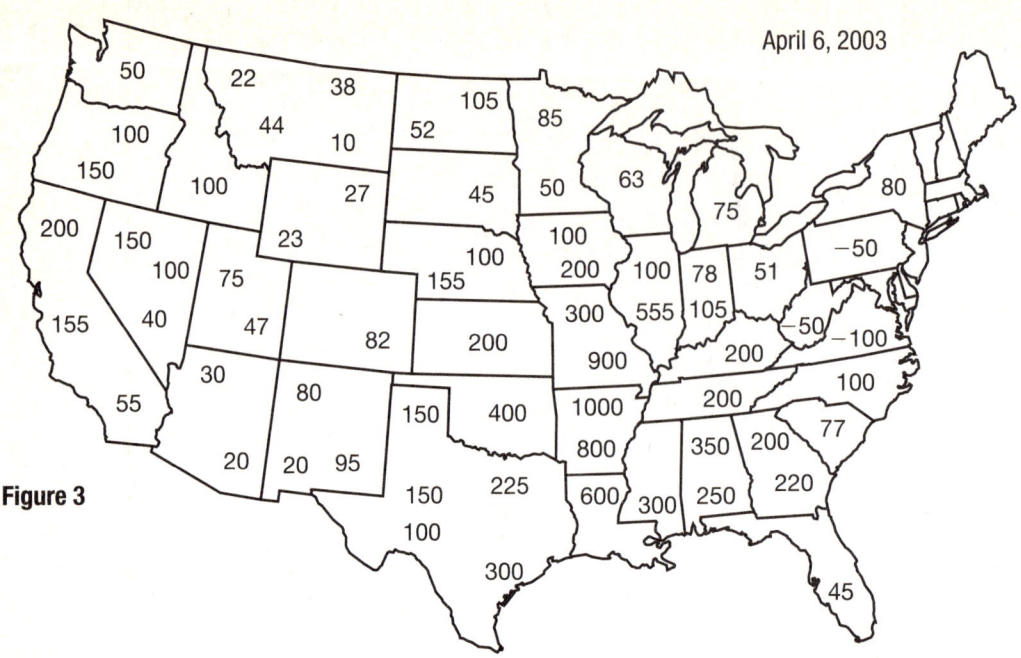

Figure 3

April 6, 2003

DATA TABLE 2

Helicity	Amount of Rotation
> 100	Some storm rotation
> 250	Enough rotation to support supercell thunderstorms and some tornadoes
> 400	Enough rotation to support dangerous tornadic thunderstorms

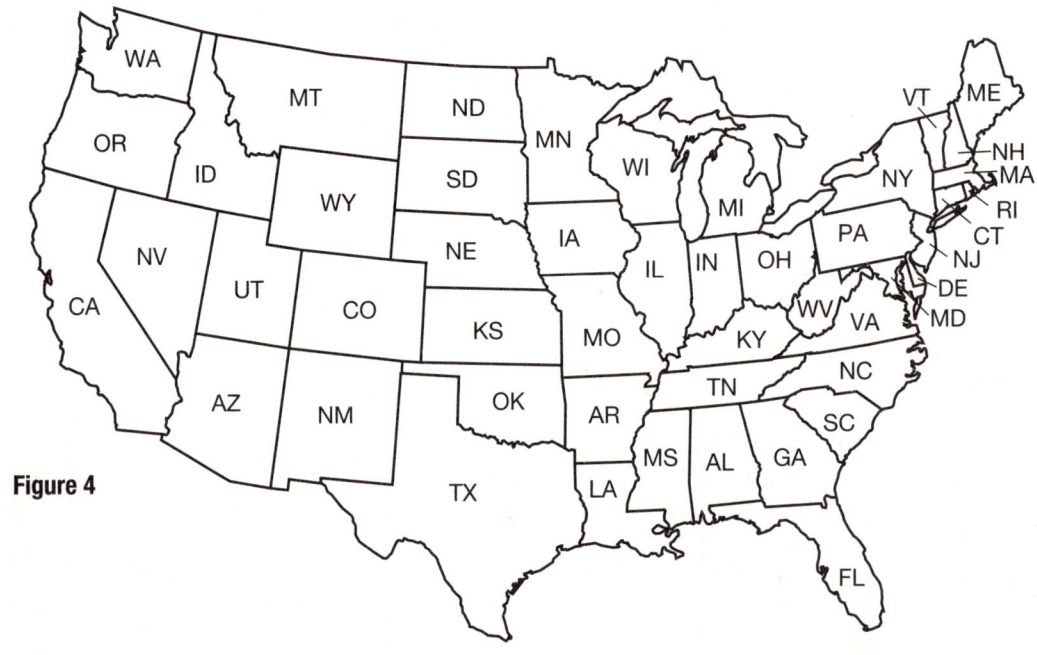

Figure 4

Name _____ Class _____ Date _____

Analysis and Conclusions

1. **Interpreting Diagrams** Considering only the dew-point data, in which states could a tornado have possibly formed on April 3, 2003?

   Texas, Arkansas, Louisiana, Mississippi, Georgia, Alabama, Florida, South Carolina

2. **Analyzing Data** Considering only the lifted index data, in which states could a tornado have possibly formed on April 3, 2003?

   Texas, Louisiana, Mississippi, Alabama

3. **Interpreting Diagrams** Based only on the storm relative helicity data, in which states could a tornado have possibly formed on April 3, 2003?

   Texas, Arkansas, Louisiana, Mississippi, Alabama, Illinois, Missouri, Oklahoma

4. **Predicting** Study your combined data, then predict which states are most likely to have experienced a tornado.

   Texas, Louisiana, Mississippi, Alabama

5. **Evaluating and Revising** What criteria did you use to make your decision?

   Sample answer: If a state had a favorable dew-point temperature but it did not have rising air or rotating air, it was eliminated. If a state had a favorable dew-point temperature but it did not have a high enough storm relative helicity value, it was eliminated.

6. **Applying Concepts** Imagine that a cold front moved across the southeastern United States on April 6, 2003. Would this front have increased or decreased the chances of tornado formation? Explain your answer.

   Cold fronts help to force up air. Rising air is associated with atmospheric instability and storm formation, so a cold front would likely increase the possibility of a tornado.

Name _____ Class _____ Date _____

Go Further

Thermodynamic indices are available on the Internet. Search the Internet for Web sites that post data about these indices. Select a state, then gather data about its dew point, lifted index, and storm relative helicity on one particular date. Predict whether the state is likely to experience a tornado.

Answers will vary depending on the date and state chosen. In general, conditions are favorable for tornado development in an area that is experiencing a dew-point temperature of at least 50°F, a lifted index of −4 or lower, and a storm relative helicity greater than 250.

Have students bookmark their Web sites. They will use these sites in Investigation 20C to create a weather station.

Name _____ Class _____ Date _____

Chapter 20   Weather Patterns and Severe Storms

Investigation 20B

Interpreting Weather Diagrams

Introduction

Every day meteorologists send weather balloons high into the atmosphere. These balloons carry weather instruments that record atmospheric conditions such as temperature, pressure, and wind speed. This information is transmitted to a computer, which then creates a printout called a Skew-T.

A **Skew-T** is a diagram that shows the condition of the atmosphere for a particular area at a particular time. It includes minor thermodynamic indices that provide valuable information to meteorologists.

In this investigation, you will use a Skew-T diagram to predict the possibility of severe weather.

Have students review the information on thunderstorms and tornadoes in their textbooks. **SKILLS FOCUS:** *Analyzing Data, Interpreting Diagrams, Predicting* **TIME REQUIRED:** *30 minutes*

Problem

How can thermodynamic indices be used to predict the possibility of a severe storm?

Pre-Lab Discussion

Read the entire investigation. Then work with a partner to answer the following questions.

1. **Controlling Variables** Why must all the data for Lake Charles be gathered at the same time and date?

   In order for the data to be compared, it must be gathered at the same time and date.

2. **Forming Definitions** What information does the Bulk Richardson Number (BRN) provide?

   The Bulk Richardson Number indicates the type of storm that may develop.

3. **Predicting** What type of weather would you expect for an area with an energy index (EI) of 3?

   no storms; clear weather

4. **Formulating Hypotheses** Using the minor thermodynamic indices as a guide, what conditions are associated with severe weather?

   The TT must be greater than 50; the SW must be greater than 300; the EI must be less than −2;

   the CAPE must be greater than 2500; the EHI must be greater than 1; and the BRN should be

   between 30 and 60.

Earth Science Lab Manual ▪ 129

Name _____ Class _____ Date _____

Procedure

1. Study the Skew-T diagram in Figure 1. The Skew-T shows minor thermodynamic indices for Lake Charles, Louisiana, on April 6, 2003.

2. Locate the six indices listed in Data Table 1 on the Skew-T diagram. Write these indices in the Data Table.

3. Use the Skew-T diagram to fill in Data Table 2 with the corresponding value for each of the six indices.

4. Based on the values on the Skew-T and Data Table 1, complete Data Table 2 with the expected weather or conditions for Lake Charles on April 6, 2003.

Tell students that meteorologists analyze the lines in the Skew-T diagram to develop forecasts. For the purposes of this investigation, the lines can be ignored. Students should instead focus on the indices listed on the right side of the diagram. Make enlarged copies of the Skew-T for students with visual impairments.

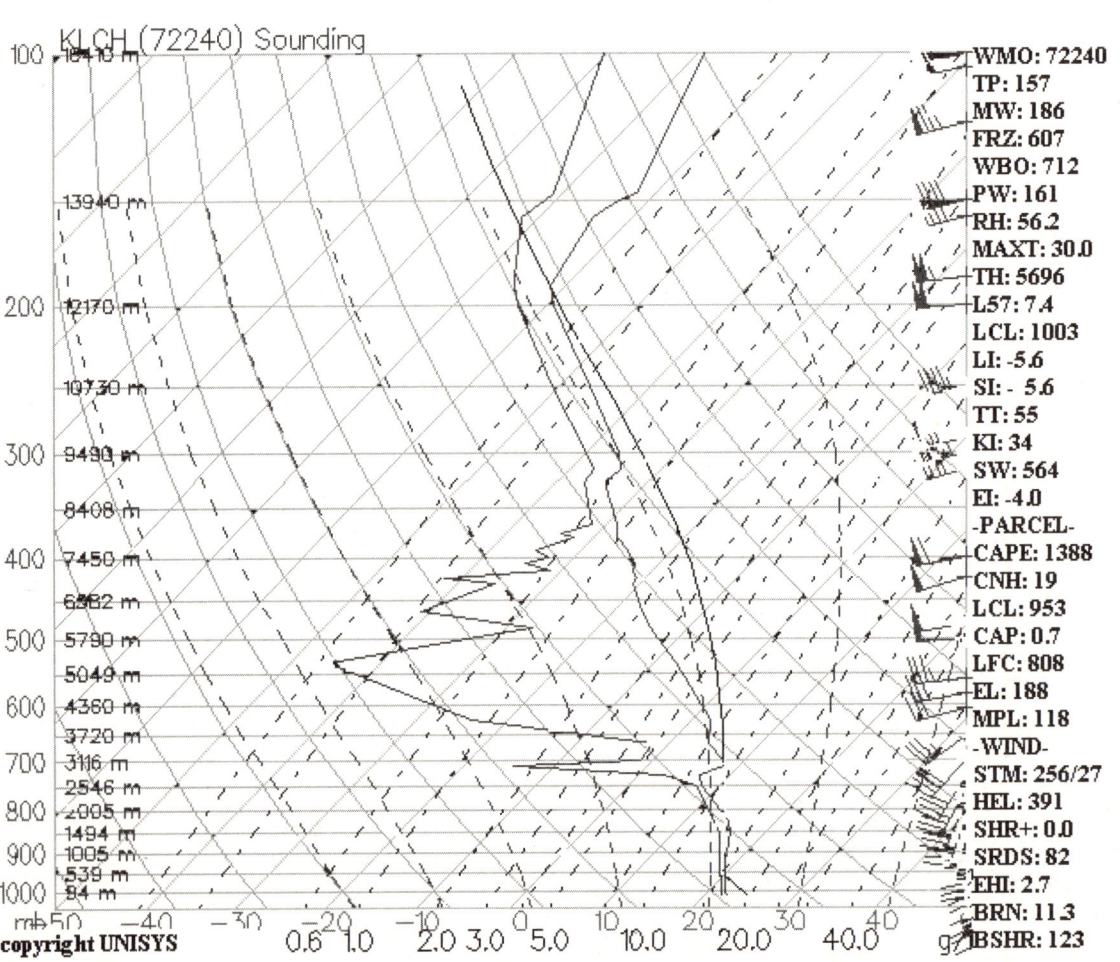

Figure 1

Name _____ Class _____ Date _____

DATA TABLE 1

TT	Type of Weather Expected
44–50	Thunderstorms are possible
51–52	Severe thunderstorms possible with isolated tornadoes
53–56	Numerous thunderstorms possible, some severe, a few tornadoes
> 56	Numerous severe thunderstorms, scattered tornadoes possible

SW	Type of Storm
> 300	Severe thunderstorms likely to occur
> 400	Tornadic thunderstorms likely to occur

EI	Type of Weather Expected
> 0	No storms expected
0 to −2	Isolated severe thunderstorms
< −2	Severe storms possible, isolated tornadoes

CAPE	Amount of Energy/Buoyancy in the Atmosphere
0–1500	Small amount
1500–2500	Moderate amount
2500–3500	Large amount

EHI	Type of Storm
< 1.0	Tornadic thunderstorms cannot be supported
> 1.0	Tornadic thunderstorms can be supported

BRN	Type of Storm That Could Develop
< 30	Storms may not form at all
30 to 60	Tornadic thunderstorms could develop
> 70	Thunderstorms with heavy rain

DATA TABLE 2   Sample data are shown.

Indices	Corresponding Values	Expected Weather/Conditions
Total Totals	55	Numerous thunderstorms possible, some severe, a few tornadoes
Severe Weather Threats	564	Tornadic thunderstorms likely to occur
Energy Index	−4.0	Severe storms possible, isolated tornadoes
Convective Available Potential Energy	1388	Small amount
Energy Helicity Index	2.7	Tornadic thunderstorms can be supported
Bulk Richardson Number	11.3	Storms may not form at all

Name _____ Class _____ Date _____

Analysis and Conclusions

1. **Interpreting Diagrams** What type of weather is associated with the Total totals (TT) for Lake Charles?

   numerous thunderstorms possible; some severe; a few tornadoes

2. **Analyzing Data** Which of the indices for Lake Charles indicate severe weather? Which indices do not?

   The indices that indicate severe weather are the TT, SW, EI, and EHI. The indices that do not indicate severe weather are the CAPE and BRN.

3. **Predicting** One can use the indices to make a rough approximation of whether atmospheric conditions could support the development of severe weather. Dividing the number of indices that indicate severe weather by the total number of indices and multiplying that number by 100 results in a percentage. Based on your data, what is the rough probability for atmospheric conditions that could support severe weather in Lake Charles? Show your work.

   > 4 indices indicating severe weather ÷ 6 indices total = 0.67 × 100 = 67 percent chance that the atmosphere could support severe weather

4. **Evaluating and Revising** Why is the method described in item 3 considered a rough approximation? Explain your answer.

   Sample answer: The Skew-T diagram includes numerous indices in addition to the six studied here. The prediction is probably not very accurate because it does not take into account all atmospheric conditions.

Go Further

As with the three major thermodynamic indices, minor indices are available on the Internet. Search the Internet for Web sites that post data about minor indices. Select a city and then gather data about the following indices: TT, SW, EI, CAPE, EHI, and BRN. Predict whether the city is likely to experience severe weather.

Answers will vary depending on the date and city chosen. In general, conditions are favorable for severe weather in a city with the following indices: TT greater than 50; SW greater than 300; EI less than −2; CAPE greater than 2500; EHI greater than 1; BRN greater than 30.

Have students bookmark the Web sites. They will use these sites in Investigation 20C to create a weather station.

Name _____ Class _____ Date _____

Chapter 20 Weather Patterns and Severe Storms

Investigation 20C

Creating a Weather Station

Have students review the information on thunderstorms and tornadoes in their textbooks. **SKILLS FOCUS:** Analyzing Data, Interpreting Diagrams, Predicting **TIME REQUIRED:** 45 minutes

Introduction
Meteorologists use numerous thermodynamic indices to help develop forecasts. Much of this information can be found on the Internet. In addition, the Internet is a good source for radar images that indicate wind speeds and storm locations. Satellite images that show differences in cloud temperatures and surface temperatures are also available on the Internet.

In this investigation, you will use data gathered from the Internet to create a weather station to make severe weather predictions. To complete this activity, you must review the thermodynamic indices discussed in the two previous labs.

Problem
How can data be organized into a weather station to make severe weather predictions?

Pre-Lab Discussion
Read the entire investigation. Review the information on thermodynamic indices in the two previous labs. Then work with a partner to answer the following questions.

1. **Forming Definitions** What does a storm relative helicity value of 250 mean?

   A value greater than 250 means there is enough rotation in the atmosphere to support supercell thunderstorms and some tornadoes.

2. **Predicting** What type of weather would you expect for an area that has a lifted index of −6?

   severe storms; large hail; possibly large tornadoes

3. **Predicting** What type of weather would you expect for an area with a severe weather threat (SW) of 460?

   Tornadic thunderstorms are likely.

4. **Designing Experiments** How will you organize your weather station?

   Sample answer: Use your Web browser to create a severe weather folder. Then move the bookmarked weather sites into the folder.

Earth Science Lab Manual ▪ 133

Name _____ Class _____ Date _____

Materials *(per pair of students)*
computer with Internet access
printer
labeled map of the United States
colored pencils

> Pair students who are computer literate with those who are less comfortable working with technology. Suggest that students access the Web sites of the National Weather Service and educational institutions for weather data.

Procedure

Part A: Creating a Weather Station

1. In Internet Explorer, go to the menu bar and click on "Favorites." If you are using Netscape, go to the menu bar and click on "Bookmarks."

2. In Internet Explorer, click on "Organize Favorites." In Netscape, click on "Manage Bookmarks."

3. In Internet Explorer, click on "Create Folder," and name the new folder "Severe Weather Station." In Netscape, click on "File" in the menu bar, then "New," and lastly "Folder."

4. Move all the Web sites you bookmarked in the two previous labs into the Severe Weather Station folder.

5. Rename the Web sites. For example, a Web site that shows a dew-point map can be called "Dew-Point Index."

6. Your station should have the following named locations: Dew-Point Index, Lifted Index, Storm Relative Helicity Index, and Skew-T Diagrams. If necessary, conduct additional research. You can add additional Web sites that show frontal boundaries and radar and satellite images.

Part B: Gathering Weather Data

7. Using a site from your weather station, print out a U.S. weather map that shows dew-point temperatures. On the map in Figure 1, draw an outline around those states that have dew-point temperatures conducive to tornado formation.

8. Repeat Step 7 using different colored pencils for the lifted index and storm relative helicity index.

9. Determine which states meet the criteria for tornado formation for all three major indices.

10. Identify several cities within these states. Print out Skew-T diagrams for these cities using the sites from your weather station. Complete the Data Table with the corresponding values for the indices for each city.

Earth Science Lab Manual ▪ **134**

Name _____ Class _____ Date _____

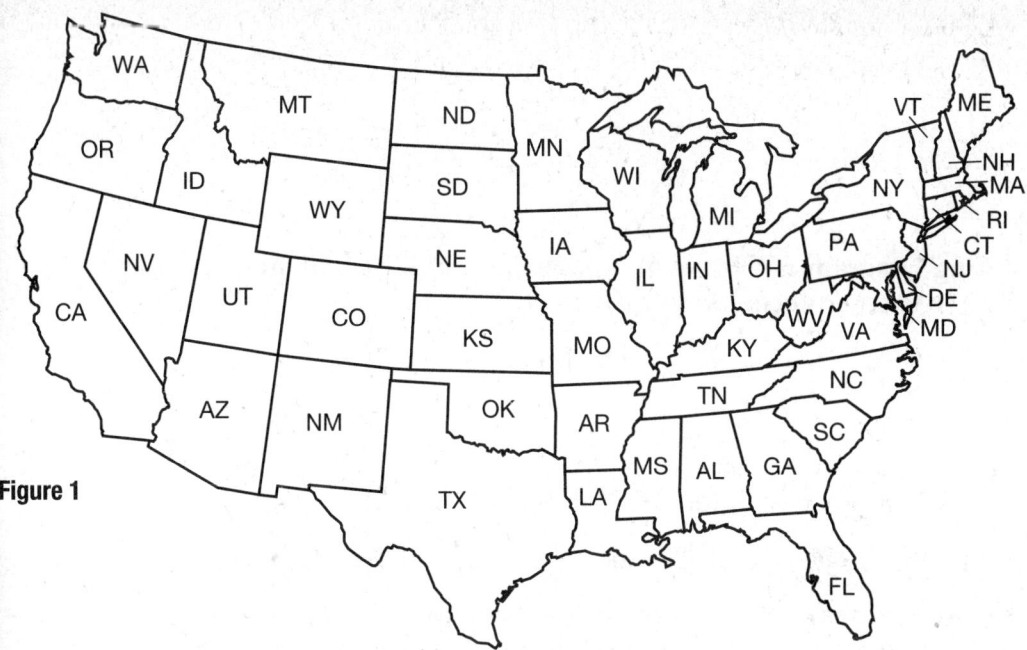

Figure 1

DATA TABLE

City	Dewpoint	L. Index	Helicity	TT	SW	EI	CAPE	EHI	BRN

Name _____ Class _____ Date _____

Analysis and Conclusions

1. **Interpreting Diagrams** Which states have dew-point temperatures that are conducive to tornado formation?

   Answers will vary. The listed states should have dew-point temperatures of at least 50°F.

2. **Interpreting Diagrams** Which states have lifted indices that are conducive to tornado formation?

   Answers will vary. The listed states should have lifted indices of at least −4.

3. **Interpreting Diagrams** Which states have storm relative helicity values that are conducive to tornado formation?

   Answers will vary. The listed states should have storm relative helicity values of at least 250.

4. **Analyzing Data** Based on your Skew-T diagrams, which of your chosen cities have at least three minor thermodynamic indices that indicate severe weather? Which cities do not have minor indices that indicate severe weather?

   Answers will vary. Conditions generally are favorable for severe weather in a city with the following indices: TT greater than 50; SW greater than 300; EI less than −2; CAPE greater than 2500; EHI greater than 1; and BRN greater than 30.

5. **Predicting** Based on your combined data, where is a tornado most likely to form?

   Answers will vary. The locations should have the following indices: dew-point of at least 50°F; lifted index of at least −4; storm relative helicity of at least 250; TT greater than 50; SW greater than 300; EI less than −2; CAPE greater than 2500; EHI greater than 1; and BRN greater than 30.

6. **Evaluating and Revising** How do you think your prediction would change if you rechecked the data in several hours?

   Sample answer: Earth's surface and atmosphere heat up as the day progresses, and they cool down as night approaches. If the data were gathered early in the day, the probability of tornadoes might increase later that afternoon. If the data were gathered in the afternoon, the probability of tornadoes might decrease toward evening.

Go Further

Use the Internet to research caps, or temperature inversions. Explain what caps are. How do caps affect storm formation?

Caps are warm layers of air high in the atmosphere that stop the upward movement of rising air. They keep storms from developing.

Name _____ Class _____ Date _____

Chapter 21   Climate   Investigation 21

Modeling the Greenhouse Effect

Have students review the information on the greenhouse effect and global warming in their textbooks.
SKILLS FOCUS: Observing, Measuring, Using Models
TIME REQUIRED: 35 minutes if bottles are already perforated; 45 minutes if students must perforate bottles before conducting the investigation

Introduction

A greenhouse is a structure whose glass or plastic panes allow light from the sun to enter the structure but also prevent heat from escaping. In a similar way, Earth's atmosphere allows solar radiation to pass through it. Some of this radiation is absorbed by Earth's surface. Gases in the atmosphere, including carbon dioxide and water vapor, also absorb some of this energy and reflect it back to Earth's surface as heat. This greenhouse effect makes our planet's surface and atmosphere warmer than they would be otherwise.

In this investigation, you will model the greenhouse effect and compare your results to the greenhouse effect caused by Earth's atmosphere.

Problem

How can you model the greenhouse effect caused by Earth's atmosphere?

Pre-Lab Discussion

Read the entire investigation. Then work with a partner to answer the following questions.

1. **Posing Questions** Write a question that summarizes the purpose of this investigation.

   Sample answer: How does Earth's atmosphere cause energy to become trapped?

   _____

2. **Controlling Variables** What is the dependent variable in this investigation?

   Temperature of the air inside the bottles is the dependent variable.

   _____

3. **Controlling Variables** What is the independent variable in this investigation?

   The independent variable is the holes in one of the bottles.

   _____

4. **Predicting** Based on the information given in the procedure, predict the outcome of this investigation.

   Answers will vary, but most students should predict that the temperature of the air inside

   the perforated bottle will be lower than the temperature inside the bottle without holes.

Earth Science Lab Manual • 137

Name _____ Class _____ Date _____

Materials *(per pair of students)*

2 clean, dry, transparent, 2-L plastic soda bottles with caps
laboratory burner
heat-resistant gloves
safety matches
large, metal knitting needle
2 identical, non-mercury, Celsius thermometers
modeling clay
direct sunlight or a gooseneck lamp with 100-W bulb
clock or watch
colored pencils

Ask each student to bring in 1 clean, clear, 2-L plastic soda bottle (and its cap) from which the label has been removed. Try to gather those bottles that are more flexible (softer). If time is a factor, perforate half of the bottles before giving them to students. As an alternative to poking holes into one of the bottles in each pair, you or students could carefully use sharp scissors or a single-edge razor blade in a safety handle to cut small pieces from the bottles. Gather or order long (~30 cm), slim thermometers with scales directly on the instruments. Students have to be able to easily get the instruments into and out of the soda bottles.

Safety

Put on safety goggles. Tie back loose hair and clothing when working with flames. Do not reach over an open flame and keep alcohol away from any open flame. Also, be careful when using matches. Use extreme care when working with heated equipment or materials to avoid burns. Be careful to avoid breakage when working with the thermometers. Note all safety symbols next to the steps in the Procedure and review the meanings of each symbol by referring to the symbol guide on page xiii.

Monitor students to insure that they follow safety procedures throughout the investigation. If necessary, review the correct procedure for lighting the lab burner.

Procedure

1. Put on safety goggles.
2. Connect the laboratory burner to the gas valve.
3. Put on heat-resistant gloves and open the valve. Carefully light the burner and properly dispose of the match.

   Remind students to allow the matches to cool before disposing of them.

4. **CAUTION:** *Carefully warm the knitting needle in the flame and use it to make 30 holes in one of the 2-L bottles.* Distribute the holes evenly around the bottle, but do not make holes around the bottom 6 cm of the bottle. Turn off the burner and put it away.
5. Lower one thermometer into each of the bottles so that the bulbs of the thermometers are at the bottoms of the bottles. Screw the caps tightly onto the bottles.
6. Use the modeling clay to secure the bottles—upside down—in an area that gets direct sunlight or under the lamp. The bottles should be about 15 cm apart. If you are using a lamp, adjust the lamp so that each bottle is the same distance—about 10 cm—from the bulb.
7. Adjust the bottles so that the thermometers are set up the same way with respect to the light source.
8. If you are using a lamp as your light source, turn it on. **CAUTION:** *Lamps can get very hot. Do not move too close to the lamps when they are in use.* Measure the initial temperature shown on each thermometer and record these values in the data table.
9. Measure and record the temperature in each bottle every 5 minutes for 30 minutes.

Earth Science Lab Manual ▪ 138

Name _____ Class _____ Date _____

Observations

DATA TABLE

Time (minutes)	Temperature (°C)	
	Bottle Without Holes	Perforated Bottle
0		
5		
10		
15		
20		
25		
30		

GRAPH

Construct a line graph of your data on the grid below. Plot time, in minutes, on the horizontal axis, and temperature, in degrees Celsius, on the vertical axis. Use a different colored pencil to connect each set of data points. Include a key that indicates which set of data is which. Give your graph an appropriate title.

Title: _____

Earth Science Lab Manual ▪ 139

Name _____ Class _____ Date _____

Analysis and Conclusions

1. **Observing** In which of the two bottles did the temperature of the air rise at a faster rate? Explain why this happened.

   The temperature of the air in the bottle without the holes rose more quickly because

   the gases inside the bottle weren't able to escape as they did in the perforated bottle.

2. **Relating Cause and Effect** In which of the bottles did the air reach the higher temperature? Why?

   The air in the closed system—the bottle without the holes—reached a higher temperature

   because the air was unable to mix with cooler air outside the bottle as it did in the perforated bottle.

3. **Inferring** Which processes of heat transfer—conduction, radiation, convection—are involved in this activity?

   convection and radiation

4. **Using Analogies** Which bottle simulates the greenhouse effect caused by Earth's atmosphere? Why?

   The closed bottle simulates the greenhouse effect; air in the bottle traps energy as heat.

5. **Evaluating and Revising** What are some of the limitations of this model of Earth's greenhouse effect?

   The closed bottle, like an actual greenhouse, prevents heat loss from both radiation and

   convection. Earth's atmosphere, on the other hand, can only prevent heat loss by radiation.

6. **Applying Concepts** How is the greenhouse effect related to global warming?

   Earth's greenhouse effect helps maintain an environment hospitable for life. Increases in

   the concentrations of certain gases may be causing Earth's surface temperatures to rise.

   This increase is called global warming.

7. **Inferring** How could you alter the lab procedure to obtain better results?

   Answers will vary. Accept all reasonable responses.

Go Further

Compare your results from this investigation with the results from the Quick Lab in your textbook. Explain the effect of the greenhouse effect on air temperature alone, air temperature above soil, and air temperature above water.

The heat-trapping ability of a greenhouse depends on the transparency of the cover and the color and composition of the surfaces inside the greenhouse.

Name _____ Class _____ Date _____

Chapter 22 Origin of Modern Astronomy   Investigation 22

Measuring the Angle of the Sun at Noon

Have students review the information on the Earth-moon-sun system in their textbooks. **SKILLS FOCUS:** Measuring, Analyzing Data, Relating Cause and Effect **TIME REQUIRED:** 30 minutes per session; four sessions total

Introduction

As Earth revolves around the sun, the orientation of Earth's axis to the sun continually changes. As a result, the location of the rising and setting sun changes throughout the year. The **altitude** is the angle above the horizon of the sun at noon. The altitude also changes throughout the year because of the orientation of Earth's axis.

In this investigation, you will indirectly observe the changing orientation of Earth's axis by measuring the altitude of the sun at noon over a period of several weeks.

Problem

How does the altitude of the sun at noon change over time?

Pre-Lab Discussion

Read the entire investigation. Then work with a partner to answer the following questions.

1. **Relating Cause and Effect** Does the sun actually move across the sky? Explain your answer.

   No, Earth is moving, not the sun. Earth moves around the sun as it rotates on its axis. The movement of Earth causes the apparent movement of the sun across the sky.

2. **Forming Definitions** What does the term *altitude* refer to in this investigation?

   Altitude is the angle above the horizon of the sun at noon.

3. **Predicting** During what time of year would you expect the shadow of the meter stick to be longest? When would it be shortest?

   It would be longest near the winter solstice. It would be shortest near the summer solstice.

Earth Science Lab Manual ▪ 141

Name _____ Class _____ Date _____

Materials *(per student pair)*
2 meter sticks
calculator
protractor

Select a flat, sunny area for students to use as an observation site. Make sure that students hold the meter stick perpendicular to the ground.

Safety

Never look directly at the sun, as it may result in eye damage. Note the safety symbols next to the steps in the Procedure and review the meaning of each symbol by referring to the symbol guide on page xiii.

Procedure

1. On a sunny afternoon—noontime is best—go outside and find a flat, sunny area. Hold a meter stick upright with one end touching the ground, as shown in Figure 1. Use the protractor to make sure that the meter stick is perpendicular to the ground.

2. Observe the shadow cast by the meter stick. Hold the meter stick steady while your partner uses the other meter stick to measure the length of the shadow. **CAUTION:** *Never look directly at the sun, as it may result in eye damage.*

3. Using your calculator, divide the height of the meter stick by the length of the shadow. Record your calculations in Data Table 1.

4. Use Data Table 2 to determine the altitude of the sun at noon. Locate the number in the table that comes closest to your calculation in Step 3. Then record the corresponding angle in Data Table 1.

5. Complete Data Table 1 with the date and time of your observation of the sun's shadow.

6. Repeat Steps 1 through 5 at exactly the same time on several different days over a period of four or five weeks. Record your results in Data Table 1.

DATA TABLE 1

Observation	Date	Time	Calculations (Height of Meter Stick Divided by Length of Shadow)	Angle of Sun
1				
2				
3				
4				

Earth Science Lab Manual

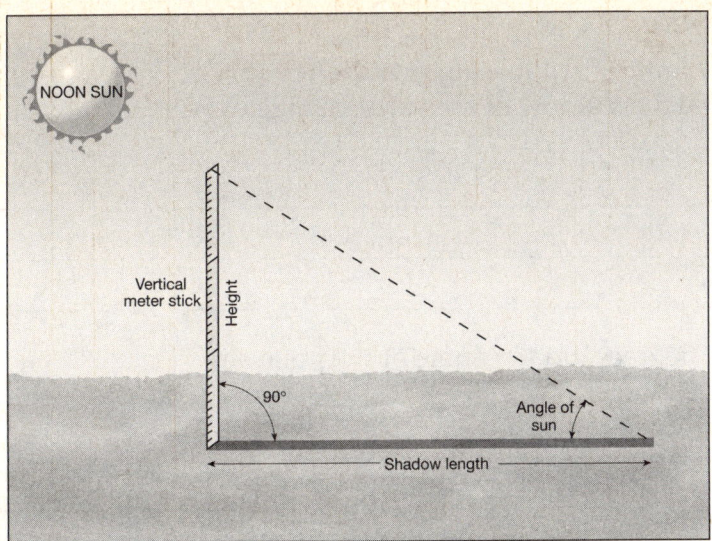

Figure 1

Data Table 2

If Height of Stick / Length of Shadow	Then Sun Angle is	If Height of Stick / Length of Shadow	Then Sun Angle is
0.2679	15°	1.235	51°
0.2867	16°	1.280	52°
0.3057	17°	1.327	53°
0.3249	18°	1.376	54°
0.3443	19°	1.428	55°
0.3640	20°	1.483	56°
0.3839	21°	1.540	57°
0.4040	22°	1.600	58°
0.4245	23°	1.664	59°
0.4452	24°	1.732	60°
0.4663	25°	1.804	61°
0.4877	26°	1.881	62°
0.5095	27°	1.963	63°
0.5317	28°	2.050	64°
0.5543	29°	2.145	65°
0.5774	30°	2.246	66°
0.6009	31°	2.356	67°
0.6249	32°	2.475	68°
0.6494	33°	2.605	69°
0.6745	34°	2.748	70°
0.7002	35°	2.904	71°
0.7265	36°	3.078	72°
0.7536	37°	3.271	73°
0.7813	38°	3.487	74°
0.8098	39°	3.732	75°
0.8391	40°	4.011	76°
0.8693	41°	4.332	77°
0.9004	42°	4.705	78°
0.9325	43°	5.145	79°
0.9657	44°	5.671	80°
1.0000	45°	6.314	81°
1.0360	46°	7.115	82°
1.0720	47°	8.144	83°
1.1110	48°	9.514	84°
1.1500	49°	11.430	85°
1.1920	50°		

Name _____ Class _____ Date _____

Analysis and Conclusions

1. **Analyzing Data** How did the altitude of the sun at noon—or the time you were able to measure the shadow of the sun—change over time?

   Answers will vary depending upon the date and location of the observations. As a general guide, the altitude of the sun at noon is greater in the summer than in the winter.

2. **Analyzing Data** How many degrees did the angle of the noon sun change over the period of your observations?

   Answers will vary depending upon the date and location of the observations.

3. **Calculating** What is the approximate average change of the angle of the noon sun per day?

   approximately 0.25°

4. **Relating Cause and Effect** Why does the angle of the noon sun change over time?

   The orientation of Earth's axis to the sun continually changes because Earth revolves around the sun. This change causes the angle of the noon sun to change over time.

5. **Designing Experiments** How might you alter the procedure to get better results?

   Answers will vary. Students may cite measuring the angle of the sun over a longer period of time or measuring the angle of the sun over two different seasons to compare how the angle of the sun changes throughout the year.

Go Further

Predict how your results would change if you repeated this investigation in six months.

Answers will vary depending upon date and location of observations. If the original measurements were taken during winter, then the length of the shadow of the meter stick would be shorter six months later during summer. The resulting angle of the sun would be greater.

Name _____ Class _____ Date _____

Chapter 23 Touring Our Solar System

Investigation 23

Exploring Orbits

Have students review the information on planetary orbits in the chapter. **SKILLS FOCUS:** *Measuring, Inferring, Comparing and Contrasting* **TIME REQUIRED:** *45 minutes*

Introduction

In 1609, the German mathematician and astronomer Johannes Kepler deciphered a major puzzle of the solar system. The strange back-and-forth motions of the planets in the sky were nearly impossible to predict until Kepler figured out the true shapes of the planetary orbits around the sun. Orbits were always believed to be circular, but Kepler used mathematics to discover they were actually **elliptical.**

An **ellipse** is an oval that is characterized by two quantities. The first quantity is the width of the ellipse, which is called the **major axis.** The second quantity is called the **eccentricity,** which is a measure of how stretched out the ellipse is. Eccentricity is defined by the distance between two mathematically determined points within the ellipse called the foci.

For planets, one focus of their orbital ellipse is the sun. The other is an empty point in space. The orbit of each planet is an ellipse, but each planet's elliptical orbit has different major axes and eccentricities. Once Kepler understood the proper nature of orbits, the movements of the planets in the sky could be predicted with precision.

In this investigation, you will draw ellipses, calculate their eccentricities, observe an interesting property of ellipses, and compare the ellipses you draw with the orbital eccentricities of Earth and other planets in the solar system.

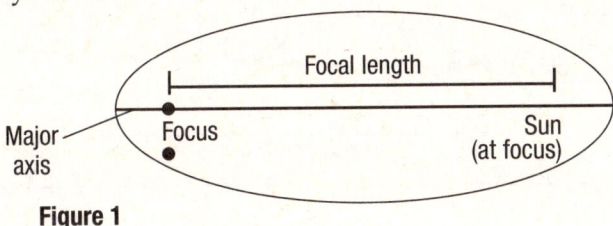

Figure 1

Problem

What do the elliptical orbits of the planets look like?

Pre-Lab Discussion

Read the entire investigation. Then work with a partner to answer the following questions.

1. **Predicting** Each planet's orbit is shaped like an ellipse. Predict whether the shapes of the planet's orbits will be more circular or more elongated.

   Students probably will predict that the orbits are elongated.
   _____
   _____
   _____

Earth Science Lab Manual ▪ **145**

Name _____ Class _____ Date _____

2. **Applying Concepts** What is the one thing that the elliptical orbits of all planets, asteroids, and most comets have in common?

   The sun is at one focus of all these elliptical orbits.

3. **Controlling Variables** What is the independent variable for the ellipses you will draw?

   The independent variable is the focal length of each ellipse.

4. **Controlling Variables** What are the dependent variables for the ellipses you will draw?

   Both the eccentricity and the major axis length are dependent variables.

5. **Designing Experiments** What purpose do the two pushpins serve in this investigation?

   The pushpins mark the foci of the ellipses.

Materials *(per pair of students)*
3 sheets of paper
heavy corrugated cardboard (~50 cm × 60 cm)
2 pushpins
metric ruler
string, 30 cm long
5 colored pencils
cellophane tape
calculator

Name _____ Class _____ Date _____

Safety Monitor students to ensure that they follow safety procedures throughout the investigation.
Be careful when handling sharp objects. Note all safety symbols next to the steps in the Procedure and review the meaning of each symbol by referring to the safety symbol guide on page xiii.

Procedure
Part A: Drawing Ellipses and Calculating Eccentricity

1. Fold a sheet of paper in half lengthwise. Flatten it out again.
2. Place the paper on the cardboard and measure 5 cm from the center of the page. Place one pushpin at the 5 cm mark. Measure 5 cm to the other side of the center of the paper and place a pushpin there. The pushpins should be 10 cm apart. **CAUTION:** *Be careful when handling the pins; they can puncture skin.*
3. Label one of the pushpins as the sun.
4. Tie the string in a loop and place it around the pins. Using one of the colored pencils, gently pull the string tight. Keep the string tight without pulling the pins out of the cardboard. Carefully drag the pencil around the pins to draw an ellipse, as shown in Figure 2.

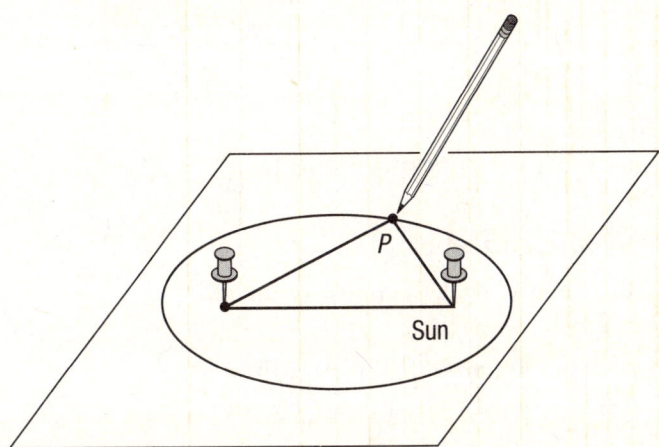

5. Using the same colored pencil, draw a circle around the pin that is not labeled as the sun. Remove the pin.
6. Use the metric ruler to measure the length of the major axis and focal length. Record these values in Data Table 1.
7. Reposition the second pin so that it is now 8.0 cm from the other pin. Repeat Steps 4 through 6, using a different colored pencil.
8. Repeat Step 7, using distances of 6.0 cm, 4.0 cm, and 2.0 cm between the pins. As the focal length for each ellipse becomes smaller, you may need to tape additional sheets of paper above and below the original sheet of paper to draw the entire ellipse.

Earth Science Lab Manual ▪ 147

Name _____ Class _____ Date _____

9. The eccentricity for each ellipse is calculated by dividing the focal length by the length of the major axis:

$$\text{eccentricity} = \frac{\text{focal length}}{\text{major axis}}$$

Calculate the eccentricity for each ellipse. Record the values in Data Table 1.

10. Label each ellipse on your diagram with its matching eccentricity.

Observations Sample data are shown.

DATA TABLE 1

Ellipse (Color)	Major Axis (cm)	Focal Length (cm)	Eccentricity
Black	20.0	10.0	0.500
Red	22.0	8.0	0.36
Blue	24.0	6.0	0.25
Green	26.0	2.0	0.15
Purple	28.0	2.0	0.071

Part B: Observing Properties of Ellipses

11. Choose one of the ellipses you made on your diagram and label the foci "A" and "B" respectively.

12. Choose a point anywhere on the ellipse and label it "C."

13. Measure the length of the lines AC and BC in centimeters. Record your measurements in Data Table 2.

14. Repeat Steps 12 and 13, placing point C at three different points on the ellipse. Carefully measure lines AC and BC and record your measurements in Data Table 2.

DATA TABLE 2

Position of C	Length of AC	Length of BC	Length of AC + BC
1			
2			
3			
4			

Name _____ Class _____ Date _____

Analysis and Conclusions

1. **Compare and Contrast** Compare the following values for planetary eccentricities to those you calculated for your ellipses. What can you state about the orbits of the various planets?

   Most planets have nearly circular orbits. The planets with greater eccentricities have more elongated orbits. The most elongated orbits are those of Pluto and Mercury. Venus and Neptune have the least elongated orbits.

Planet	Eccentricity
Mercury	0.206
Venus	0.007
Earth	0.017
Mars	0.093
Jupiter	0.048
Saturn	0.056
Uranus	0.047
Neptune	0.009
Pluto	0.250

2. **Inferring** What shape would you make if both pushpins were placed at a single central point? What would be the focal length and eccentricity of this shape?

   The resulting shape would be a circle with focal length and eccentricity of zero.

3. **Observing** What did you discover in Part B about the sums of lines AC and BC for your ellipse? Generalize your findings as a "basic law of ellipses."

   Students should have discovered that all the sums were equal. Sample generalization: A basic law of ellipses is that the sum of the lengths of two lines drawn from each foci to a single point on the ellipse will be equal regardless of where the point on the ellipse is located.

Earth Science Lab Manual ▪ 149

Name _____ Class _____ Date _____

4. **Drawing Conclusions** Did your results from this activity confirm your original prediction? Explain why or why not.

   Students will probably be surprised by how nearly circular the planetary orbits are. If they predicted elongated ellipses, their predictions were not confirmed. Even the orbits of Pluto and Mercury are only slightly flattened.

5. **Inferring** What body in the solar system do you think is one focus of the moon's orbit?

   Because the moon orbits Earth, Earth must be one focus of the moon's orbital ellipse.

6. **Inferring** How would you modify this activity to offer a better sense of planetary orbits?

   Students' answers will vary.

Go Further

Research the orbits of smaller bodies in the solar system such as asteroids or comets. Use the materials from this investigation and researched values for the major axis and eccentricity to produce drawings of the orbits of these objects. Include in your report your drawings and all values used.

Review student information before students begin drawing the orbits to be sure that eccentricities are not too small (not much less than 0.10) nor too large (greater than 0.90).

Earth Science Lab Manual ▪ 150

Name _____ Class _____ Date _____

Chapter 24 Studying the Sun   Investigation 24

Measuring the Diameter of the Sun

Refer students to the section on the sun in their textbooks. **SKILLS FOCUS:** Observing, Measuring, Calculating **TIME REQUIRED:** 45 minutes

Introduction

The sun is approximately 150,000,000 km from Earth. To understand how far away this is, consider the fact that light travels approximately 300,000 km/s. At this speed, it takes the light from the sun a little more than eight minutes to reach Earth.

Even though the sun is extremely far away, it is still possible to make an approximate measurement of its size. The sun's diameter can be estimated by making two simple measurements and then solving a proportion problem.

$$\frac{\text{diameter of sun}}{\text{distance to sun}} = \frac{\text{diameter of sun's image}}{\text{distance between two cards}}$$

If you can determine three of the terms in the proportion problem, the fourth term, the diameter of the sun, can be solved mathematically.

In this investigation, you will construct a simple device and use it to collect data that will enable you to calculate the diameter of the sun.

Problem

What is the diameter of the sun and how can it be determined?

Pre-Lab Discussion

Read the entire investigation. Then work with a partner to answer the following questions.

1. **Inferring** What is the purpose of this investigation?
   The purpose is to estimate the diameter of the sun.

2. **Calculating** To prepare for this calculation, solve for x in the following proportion problems.

   a. $\dfrac{x}{5} = \dfrac{100,000}{20}$

   $x = 25,000$

   b. $\dfrac{x}{5} = \dfrac{200,000}{50}$

   $x = 20,000$

Name _____ Class _____ Date _____

3. Inferring Why is it important to never look directly at the sun?

Looking directly at the sun can be harmful to the eyes. It can cause blindness because of the high-energy ultraviolet radiation that the sun emits.

4. Applying Concepts How are the cards used in this investigation? How are the cards and the proportional relationships useful for determining the diameter of the sun?

Student answers will vary, but they should indicate that the cards are used to get an image of the sun so it can be measured. The image of the sun on the card is a scaled-down version of the actual sun, so the relationship of size to distance for the sun also occurs in the reduced image, although the values are smaller.

5. Predicting Do you think your calculation of the sun's diameter will be completely accurate? Explain your answer.

No, it will not be completely accurate because all measurements have some error. We are measuring an image of the sun on an extremely small scale and using that measurement to estimate its diameter.

Materials *(per group)*
2 index cards (10 cm × 15 cm)
metric ruler
drawing compass
tape
meter stick
scissors

Safety
Be careful when handling sharp instruments. **CAUTION:** *Never look directly at the sun.* Note all safety symbols next to the steps in the Procedure and review the meaning of each symbol by referring to the symbol guide on page xiii.

Procedure
Part A: Measuring Distances and Calculating Ratios
1. Measure the base of each of the two triangles in Figure 1. Record your measurements in Data Table 1.
2. Measure the altitude (distance from tip to base) of each of the two triangles in Figure 1. Record your measurements in Data Table 1.
3. Determine the ratio between the base of the large triangle and the base of the small triangle. Record this ratio in Data Table 1.

4. Determine the ratio between the altitude of Triangle 1 and the altitude of Triangle 2. Record this ratio in Data Table 1.

5. Think about how these two ratios compare. In Part B of this lab, you will use a similar procedure to determine the diameter of the sun.
   - The base of Triangle 2 will represent the diameter of the image of the sun on a card.
   - The altitude of Triangle 2 will represent the distance between the two cards in the device you will construct.
   - The altitude of Triangle 1 will represent the distance from Earth to the sun.
   - The base of Triangle 1 will represent the diameter of the sun, which you will determine.

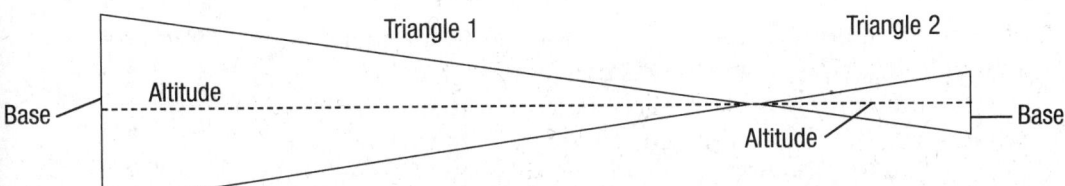

Figure 1

Part B: Determining the Diameter of the Sun

6. Using the scissors, cut I-shaped slits in each card in the positions shown in Figure 2. The meter stick should be able to slide through the slits, but the slits should be small enough so that the meter stick fits snugly. **CAUTION:** *Be careful when handling sharp instruments.*

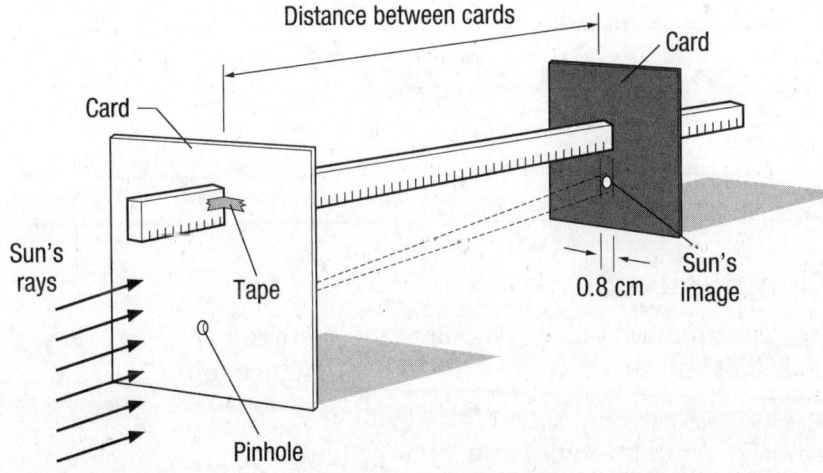

Figure 2

7. With the tip of the compass, punch a round pinhole in one of the cards in the position shown in Figure 2. Tape this card to the meter stick at the 5-cm mark so that it is perpendicular to the meter stick. **CAUTION:** *Be careful when handling sharp instruments.*

Name _____ Class _____ Date _____

8. On the other card, draw two parallel lines exactly 0.8 cm (8 mm) apart directly below the slit, as shown in Figure 2. Slide this card onto the meter stick. Do not tape this card to the meter stick.

9. While outdoors on a sunny day, position the meter stick so that the taped card is directly facing the sun. Position the meter stick until it casts a shadow over the movable card. **CAUTION:** *Never look directly at the sun.*

10. You should be able to see a circle of light on the movable card caused by the sun's rays passing through the pinhole on the first card. If you do not see the circle of light, continue to adjust the position of the meter stick until you see the circle.

11. The circle of light on the second card is an image of the sun. Slide the movable card until the image of the sun fits exactly between the two parallel lines you drew earlier.

12. Make sure that both cards are perpendicular to the meter stick. You will know they are perpendicular when the circle of light, the sun's image, is brightest and sharpest and as close to a circle as possible. Tape the second card in place. Measure the distance between the two cards. Record your measurement in Data Table 2.

For best results, the investigation should be conducted in bright sunshine near midday, when the sun's rays are most direct. For a sharper image of the sun, tape an empty cardboard toilet-paper tube to the second card so that the image appears inside the tube on the card. Do not attempt this investigation if it is cloudy or windy outside.

Observations Sample data are shown.

DATA TABLE 1

	Base	Altitude
Triangle 1 (large triangle)	2.55 cm	8.90 cm
Triangle 2 (small triangle)	0.85 cm	3.00 cm
Ratio (large:small)	3.00:1	2.97:1

DATA TABLE 2

Distance between two cards	80 cm
Diameter of sun's image	0.8 cm

Analysis and Conclusions

1. **Calculating** Using the equation below, calculate the diameter of the sun. Use 150,000,000 km (or 1.5×10^8 km) as the distance to the sun. Show your work.

$$\frac{\text{diameter of sun (km)}}{\text{distance to sun (km)}} = \frac{\text{diameter of sun's image (cm)}}{\text{distance between two cards (cm)}}$$

Sample answer: $\frac{x \text{ km}}{150{,}000{,}000 \text{ km}} = \frac{0.8 \text{ cm}}{80 \text{ cm}}$

$x = 1{,}500{,}000$ km

Name _____ Class _____ Date _____

2. Calculating The actual diameter of the sun is 1,391,000 km. Using the equation below, determine the percentage error in your calculated value for the sun's diameter. Show your work.

$$\text{percentage error} = \frac{\text{difference between your value and the correct value}}{\text{correct value}} \times 100$$

> Sample answer: % error = [(1,500,000 − 1,391,000)/1,391,000] × 100 = 7.8%

3. Analyzing Data What could account for the difference in your calculation of the sun's diameter and the actual diameter of the sun?

There may have been error involved in accurately measuring the diameter of the sun's image or the distance between the cards. Also, an average value for the distance between the sun and Earth was used. In fact, the distance to the sun varies from a maximum of 152,100,000 km to a minimum of 147,100,000 km.

4. Applying Concepts How might the technique used in this investigation be useful in making other astronomical measurements?

With a light-gathering instrument such as a telescope, the diameters of planets or the moon can be measured.

5. Relating Cause and Effect How might clouds in the sky affect the accuracy of your measurement in this investigation?

If a cloud partially covers the sun, the image might appear smaller in diameter than it normally would. Total cloud cover would make the image too faint to be seen and measured properly.

Earth Science Lab Manual

Name _____ Class _____ Date _____

Go Further

A sunspot moves along the sun's equator. If the sunspot takes 12.5 days to move from one side of the sun to the other, use the steps below to calculate how fast the sunspot is moving.

1. Using the actual value for the diameter of the sun and the formula below, calculate the circumference of the sun. The value of π (pi) is approximately 3.14.

   circumference = $\pi \times$ diameter

   > C = 3.14 × 1,391,000 km = 4,368,000 km

2. The sunspot moved only halfway around the sun, so to calculate the distance it traveled in 12.5 days, divide the value for the circumference by 2.

   > Distance = C ÷ 2 = 4,368,000 km ÷ 2 = 2,184,000 km

3. To calculate the distance traveled by the sunspot in one day, divide the distance you calculated in Step 2 by 12.5.

   > Speed = distance/time = 2,184,000 km/12.5 days = 175,000 km/day

4. Explain why this value is also the speed at which the sun's surface is moving at the equator.

   > The sunspot is part of the sun's surface, so it moves as the sun moves. Therefore, the speed of the sunspot is the same as the speed at which the sun's surface is moving at the equator.

Name _____ Class _____ Date _____

Chapter 25 Beyond Our Solar System   Investigation 25

Modeling the Rotation of Neutron Stars

Have students review the information on neutron stars and supernovae in their textbooks. **SKILLS FOCUS:** Observing, Measuring, Using Models **TIME REQUIRED:** 45 minutes

Introduction

Any spinning object will spin faster if it contracts. This concept is called **conservation of angular momentum.** You can see it in action when figure skaters execute spins. They slowly pull in their arms to make themselves twirl faster and faster. By pulling in their arms, they pull their mass closer to their rotational axis.

The same thing happens to stars when they collapse. During a supernova, the core of a star collapses into a very hot neutron star about 20 km in diameter. Some of the star's mass is lost in the explosion, but the mass that remains is tightly compressed around the rotational axis. As the star collapses, it will rotate faster for the same reason ice skaters rotate faster as they pull in their arms. The collapsing star can rotate at a speed of up to 1000 rotations per second.

In this investigation, you will use a rotating square of cardboard and masses to model the effects of mass contraction on rates of rotation.

Problem

What causes neutron stars to rotate so rapidly?

Pre-Lab Discussion

Read the entire investigation. Then work with a partner to answer the following questions.

1. **Formulating Hypotheses** Write a hypothesis that states what you expect will happen to the rate of spin when you decrease the distance between the masses on the cardboard square.

   Students should hypothesize that the rate of spin will increase when the weights are closer to the axis.

2. **Controlling Variables** What is the independent variable in this investigation?

   The distance between the weights—their distance from the axis—is the independent variable.

3. **Controlling Variables** What is the dependent variable in this investigation?

   The dependent variable is the speed of rotation.

Name _____ Class _____ Date _____

4. **Observing** After a supergiant becomes a neutron star, what is its comparative size and mass? What happens to its rotation speed?

   The size and mass will decrease. The rotational speed will increase.

5. **Inferring** How will this investigation help to answer the question of why neutron stars rotate so rapidly?

   The rotating cardboard and attached masses will model the effect of mass distribution on rotation rate. This model will demonstrate how the location of the mass will affect the rotation.

Materials *(per group)*
pencil with eraser
square of stiff cardboard, at least 35 cm wide
pushpin
masking tape
large, metal knitting needle
2 identical small masses
metric ruler
clock or watch with second hand

Provide small stones or balance masses. As the mass of the objects increases, it may pull the pushpin out of the eraser while rotating. Oddly shaped masses can be difficult to tape down. A nickel has a mass of 5 g. Students can use multiple nickels to increase the mass.

Students might have to experiment with the masses to find which work best for their cardboard. Corrugated cardboard works the best. **Note:** *Bigger cardboard squares will show a bigger difference in rotation when mass distribution is changed.*

Monitor students to ensure that they follow safety procedures throughout the investigation.

Safety
Put on safety goggles. Be careful when handling sharp objects. Make sure that your weights are securely attached so they will not fly off and hit anyone. Note all safety alert symbols next to the steps in the Procedure and review the meaning of each symbol by referring to the Safety Symbols on page xiii.

Procedure

1. Draw lines along the two diagonals of the cardboard square, as shown in Figure 1. Punch a hole through the center of the cardboard with the pushpin.
   CAUTION: *Be careful when handling sharp objects.*

2. Push the pushpin through the cardboard and into the eraser end of the pencil. The cardboard should spin freely. If it sticks, use the pushpin to widen the hole.

3. Attach a piece of masking tape to one corner of the square to serve as a marker. Measure 30 cm from the center of the cardboard to the corner and use masking tape to attach one of the masses, as shown in Figure 1. Measure 30 cm from the center to the other corner and use masking tape to attach the other mass. The masses should both lie along one of the two diagonals that pass through the center. The card will be balanced because the masses are the same distance from the center.

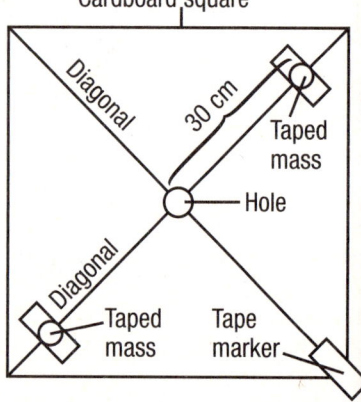

Figure 1

Earth Science Lab Manual • 158

4. Hold the pencil in front of you and perpendicular to your body so that the square of cardboard rotates below eye level and parallel to your body. Try to give each spin of the square the same force. **CAUTION:** *Make sure that masses stay attached and do not fly off and hit anyone.*

5. Spin the square. Watch the taped corner as it goes by and count the number of times it passes in 5 seconds. If the square slows down during this time, adjust the pushpin so that it turns with less friction. Record the number of rotations in the Data Table.

6. Move the masses inward toward the center so that each is 20 cm from the center and tape them down. Spin the cardboard and count the number of rotations in 5 seconds. Make sure that you apply the same force you did before. Record the number of rotations in the Data Table.

7. Repeat Step 6, but tape the masses so that each is only 10 cm from the center. Record the number of rotations in 5 seconds in the Data Table.

Observations

DATA TABLE Sample data are given.

Distance of Masses from Center of Square	Number of Rotations (spins)
30 cm	10 (for a 25-g mass)
20 cm	13
10 cm	22

Analysis and Conclusions

1. **Calculate** Calculate the rotation rate for each trial by dividing the number of spins by 5 seconds.

   Results will vary, but the rates should increase as the distance between the masses decreases.

2. **Making Judgments** Did any unexpected variable seem to affect the outcome of your investigation?

   Students should cite something other than the independent variable. They may cite the difficulties they had in applying the same force to the spin each time.

3. **Inferring** Why was it important that the masses were the same distance from the axis for each trial?

   If the masses were not at an equal distance from the center, they would not have been balanced. This lack of balance would lead to an inaccurate assessment of the responding variable.

Name _____ Class _____ Date _____

4. Infer Explain why the rate of rotation of a neutron star is so great compared to that of the star from which it formed.

A neutron star is much smaller, but it retains a great deal of mass. The size of the star decreases dramatically. The effect of its decreasing mass is much less than the effect of its decreasing size. Both effects cause the rotation rate to increase.

5. Evaluating and Revising How could you improve this model to more accurately represent the concept of conservation of angular momentum as it pertains to neutron stars?

Students might suggest that a small motor could be used to spin the pencil to give more consistent results. Some students will note that stars behave like fluids, so stars might show different results than the solids used in the model. Suggestions will vary as to how to address this. Some students will note that the loss of mass stars undergo during the supernova stage is not represented in this model. They might suggest using smaller masses as the distance between the masses is reduced with each trial.

Go Further

What happens to the rotation rate when mass decreases? Repeat Step 6 two more times, but for the first trial, use masses that are about one-half the mass of the initial masses used. For the second trial, use masses that are one-fourth the mass of the initial masses used. Create a data table in which to record your results.

Students could compare the number of rotations of the cardboard square for three different masses located the same distance from the center. The tape is sometimes difficult for students to see when the square is spinning very fast, so there may be a wide variance from the theoretical number of spins.

DATA TABLE

Mass of Objects Taped at 20 cm	Number of Rotations (Spins) from Center of Cardboard Square
25 g	13
15 g	22
5 g	38

Name _____ Class _____ Date _____

Chapter 1 Introduction to Earth Science   **Exploration Lab**

Determining Latitude and Longitude

See the *Earth Science Teacher's Edition* for more information.

Using maps and globes to find places and features on Earth's surface is an essential skill required of all Earth scientists. The grid that is formed by lines of latitude and longitude form the basis for locating points on Earth. Latitude lines indicate north-south distance, and longitude lines indicate east-west distance. Degrees are used to mark latitude and longitude distances on Earth's surface. Degrees can be divided into sixty equal parts called minutes ('), and a minute of angle can be divided into sixty parts called seconds ("). Thus, 31°10'20" means 31 degrees, 10 minutes, and 20 seconds. This exercise will introduce you to the systems used for determining location on Earth.

Problem How are latitude and longitude calculated, and how do they indicate a particular location on the globe?

Materials
- globe
- protractor
- ruler
- world map

Skills Interpreting, Measuring, Inferring

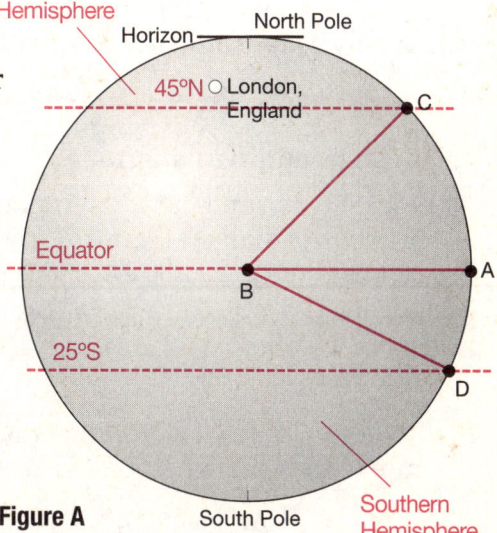

Figure A

Procedure

Part A: Determining Latitude

1. Figure A represents Earth, with point B its center. Locate the equator on the globe. Sketch and label the equator on Figure A. Label the Northern Hemisphere and Southern Hemisphere on Figure A. *See Figure A above.*

2. On Figure A, make an angle by drawing a line from point A on the equator to point B (the center of Earth). Then extend the line from point B to point C in the Northern Hemisphere. The angle you have drawn (∠ABC) is 45°. By definition of latitude, point C is located at 45°N latitude. *See Figure A above.*

3. Draw a line on Figure A through point C that is also parallel to the equator. What is the latitude at all points on this line? Record this number on the line you draw. *The latitude is 45°N. See Figure A above.*

4. Draw a line on Figure A from point D to point B. Using a protractor, measure ∠ABD on your paper. Then draw a line parallel to the equator that also goes through point D. Label the line with its proper latitude. *The latitude is 25°S. See Figure A above.*

5. How many degrees of latitude separate the latitude lines (or parallels) on the globe that you are using? Record the degrees of latitude.

Answers will vary depending on the globe used.

Earth Science Lab Manual ▪ 161

Name _____ Class _____ Date _____

6. Refer to Figure B. Determine the latitude for each point A–F. Be sure to indicate whether it is north or south of the equator and include the word "latitude." Record these numbers.

   A: 30°N latitude, B: 5°S latitude, C: 55°N latitude, D: 35°S latitude, E: 0° latitude, F: 20°N latitude

7. Use a globe or map to locate the cities listed below. Record their latitude to the nearest degree.

   A. Moscow, Russia 56°N 37°E

   B. Durban, South Africa 39°S 31°E

   C. Your home city Answers will vary depending on where students live.

8. Use the globe or map to find the name of a city or feature that is equally as far south of the equator as your home city is north.

   Answers will vary depending on where students live.

Part B: Determining Longitude

9. Locate the prime meridian on Figure C. Sketch and label the prime meridian on Figure C. Label the Eastern and Western Hemispheres. See Figure C on next page.

10. How many degrees of longitude separate each meridian on your globe?

    Answers will vary depending on where students live.

11. Refer to Figure C. Determine the longitude for each point A–F. Be sure to indicate whether it is east or west of the prime meridian.

    A: 30°E, B: 20°W, C: 45°E, D: 75°W, E: 65°E, F: 68°W

12. Use the globe or map to give the name of a city or feature that is equally as far east of the prime meridian as your home city is west.

    Answers will vary depending on where students live.

Analyze and Conclude

1. **Applying Concepts** What is the maximum number of 1 degree longitude or latitude lines that can be drawn on a globe?

   360 longitude lines and 180 latitude lines

2. **Comparing and Contrasting** Why do longitude lines converge while latitude lines do not?

   Lines of latitude all run parallel to each other. Because lines of longitude run north and south, they follow the curvature of Earth and meet at the poles.

Name _____ Class _____ Date _____

3. **Thinking Critically** Amelia Earhart, her flight engineer, and her plane are believed to have been lost somewhere over the Pacific Ocean. It is now thought that the coordinates that she was given for her fuel stop at Howley Island in the Pacific Ocean were wrong. Knowing what you do about how latitude and longitude coordinates are written, why would a wrong number have been so catastrophic for her?

<u>Earth is very large. Even a mistake of a minute or seconds of a degree may have set</u>

<u>Earhart in the wrong direction, unable to spot a small island in the ocean.</u>

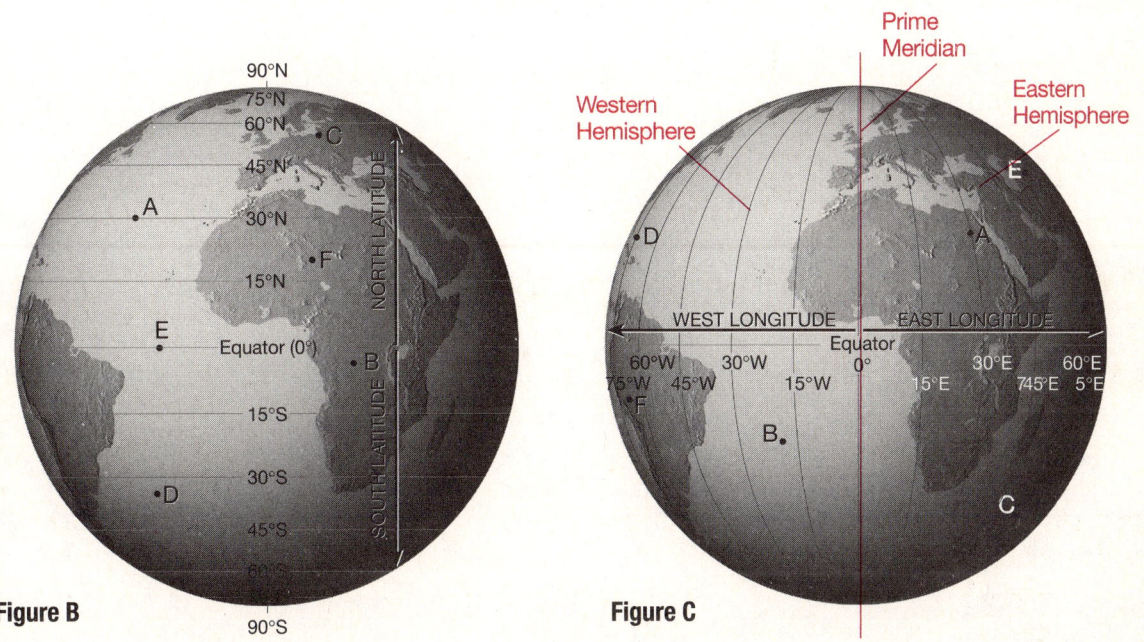

Figure B

Figure C

Earth Science Lab Manual ▪ 163

Name _____ Class _____ Date _____

Chapter 2 Minerals   Exploration Lab

Mineral Identification

See the *Earth Science Teacher's Edition* for more information.

Most minerals can be easily identified by using the properties discussed in this chapter. In this lab, you will use what you have learned about mineral properties and the table in the DataBank to identify some common rock-forming minerals. In Chapter 3, you will learn about rocks, which are mixtures of one or more minerals. Being able to identify minerals will enable you to understand more about the processes that form and change the rocks at and beneath Earth's surface.

Problem How can you use simple tests and tools to identify common minerals?

Materials
- mineral samples
- hand lens
- streak plate
- copper penny
- steel knife blade
- glass plate
- piece of quartz
- dilute hydrochloric acid
- magnet
- hammer
- 50-mL graduated cylinder
- tap water
- balance
- thin thread
- scissors
- paper or cloth towels
- Resource 16 in the DataBank

Skills Observing, Comparing and Contrasting, Measuring

Procedure

Part A: Color and Luster

1. Examine each mineral sample with and without the hand lens. Examine both the central part of each mineral as well as the edges of the samples.
2. Record the color and luster of each sample in the Data Table.

Compare students' data and observations with the information provided in Resource 16 in the DataBank.

DATA TABLE

Mineral Number	Color	Luster	Streak	Relative Hardness	Cleavage/ Fracture	Density				Other Properties
						m	V_1	V_2	d	
1										
2										
3										
4										
5										
6										
7										
8										

Part B: Streak and Hardness

3. To determine the streak of a mineral, gently drag it across the streak plate and observe the color of the powdered mineral. If a mineral is harder than the streak plate (H = 7), it will not produce a streak.

Name _____ Class _____ Date _____

4. Record the streak color for each mineral in the Data Table.
5. Use your fingernail, the penny, the glass plate, the knife blade, and the piece of quartz to test the hardness of each mineral. Remember that if a mineral scratches an object, the mineral is harder than the object. If an object scratches a mineral, the mineral is softer than the object.
6. Record the hardness values for each sample in the Data Table.

Part C: Cleavage and Fracture

7. Gently strike one of the mineral samples with a hammer. **CAUTION:** *Be sure to put on your goggles. Make sure that everyone is out of the way of flying pieces.*
8. Observe the broken mineral pieces. Does the mineral cleave or fracture? Remember that cleavage is breakage along flat, even surfaces, and fracture is uneven breakage. Record your observations in the Data Table.
9. Repeat Steps 7 and 8 for the other minerals. Record your observations for each mineral in the Data Table.

Part D: Density

10. Using a balance, determine the mass of your mineral samples. Record the mass in the first column (m) under Density.
11. Cut a piece of thread about 20 cm long. Tie a small piece of one mineral sample to one end of the thread.
12. Securely tie the other end of the thread to a pencil or pen.
13. Fill the graduated cylinder about half-full with water. Record the exact volume of the water in the second column (V_1) under Density.
14. Lower the mineral into the graduated cylinder. Read the volume of the water. Record the volume in the third column (V_2).
15. Calculate the density of the mineral using the following equation:

$$\frac{\text{mass}_1}{\text{volume}_2 - \text{volume}_1}$$

Record this value in the fourth column (d). Repeat Steps 10–15 for the other minerals. Record the densities in the Data Table.

Part E: Other Properties

16. Use the magnet to determine if any of the minerals are magnetic. Record your observations in the Data Table under Other Properties.
17. Place the transparent minerals over a word on this page to see if any have the property of double refraction. If a mineral has this property, you will see two sets of the word. Record your observations in the last column under Other Properties.
18. Compare the feel of the minerals. In the Data Table, note any differences in the last column.

Name _____ Class _____ Date _____

19. Carefully place one or two drops of dilute hydrochloric acid on each mineral. Record your observations in the last column. When you are finished with this test, wash the minerals well with tap water to rinse away the acid. **CAUTION:** *Always be careful when working with acids.*

Analyze and Conclude

1. **Identifying** Use the data and Resource 16 in the DataBank to identify each of the minerals tested.

   Most students should be able to correctly identify at least half of the minerals provided.

2. **Evaluating** Which of the properties did you find most useful? Least useful? Give reasons for your answers.

   Answers will vary, but might include color and luster as the two least useful properties and density, streak, hardness, and cleavage/fracture as most useful.

3. **Comparing and Contrasting** In general, how did the minerals with metallic luster differ from those with nonmetallic luster?

   Minerals with metallic luster tend to have higher densities than minerals with nonmetallic luster.

4. **Classifying** Classify your minerals into at least three groups based on your observations. How does your classification scheme differ from those of at least two other students?

   Students' classification schemes will probably differ but might be based on similarities and differences in relative hardness, density, luster, cleavage/fracture, or color.

Earth Science Lab Manual

Name _____ Class _____ Date _____

Chapter 3   Rocks   Exploration Lab

Rock Identification

See the *Earth Science Teacher's Edition* for more information.

Most rocks can be easily identified by texture and composition. In this lab, you will use what you have learned about rocks as well as the information on minerals from Chapter 2 to identify some common rocks.

Problem How can you use composition and texture to identify common rocks?

Materials
- rock samples
- hand lens
- pocket knife
- dilute hydrochloric acid
- colored pencils
- Resources 16 and 17 in the DataBank

Skills Observing, Comparing and Contrasting, Measuring

Procedure

1. Use the Data Table to record your observations. Add any other columns that you think might be useful.
2. Examine each rock specimen with and without the hand lens. Determine and record the overall color of each rock in the Data Table.
3. Try to identify all of the minerals in each rock, using the information in Resources 16 and 17 in the DataBank. Record your observations in the Data Table.
4. Determine and record the presence of any organic matter in any of the samples.
5. Observe the relationships among the minerals in each rock to determine texture. Refer to Resources 16 and 17 in the DataBank if necessary. Record your observations.
6. Note and record any other unique observations of the samples. **CAUTION:** *Always be careful when working with acids.*
7. In the Data Table, make and color a detailed sketch of each sample.
8. Identify each sample as being an igneous rock, a sedimentary rock, or a metamorphic rock.
9. Name each sample. Use the photographs in this chapter and Resources 16 and 17 in the DataBank if necessary.

Name _____ Class _____ Date _____

DATA TABLE Compare students' data and observations with the information in the tables in the DataBank.

Rock Number	Overall Color	Composition	Texture	Sketch	Rock Type	Rock Name
1.						
2.						
3.						
4.						
5.						

Analyze and Conclude

1. **Evaluating** Which of the rock identification characteristics did you find most useful? Which of the characteristics did you find least useful? Give reasons for your answers.

   Sample answer: Texture was more useful than composition—it was easier to determine texture than to determine composition.

2. **Comparing and Contrasting** How did identifying rocks compare with the mineral identification lab you did in Chapter 2? How is identifying rocks different from identifying the minerals that compose the rocks?

   Sample answer: The processes are similar because both involved determining the identity of a material using specific properties. The processes differ in that texture and composition are not standard properties used in mineral identification.

3. **Applying Concepts** Match the metamorphic rocks with their probable parent rocks.

   Bituminous coal is the parent rock of anthracite; sandstone the parent rock of quartzite; limestone the parent rock of marble; granite the parent rock of gneiss; and shale the parent rock of slate.

4. **Applying Concepts** Choose two pairs of rocks used in this investigation. Write a brief description for each pair that explains how one rock can be changed into the other. Refer to the diagram of the rock cycle in Chapter 3.

   Sample answer: Pressure and heat can change shale, a sedimentary rock, into slate, a metamorphic rock. Students' answers should follow the paths described in the rock cycle shown in Chapter 3.

Name _____ Class _____ Date _____

Chapter 4   Earth's Resources   Application Lab

Finding the Product that Best Conserves Resources

See the *Earth Science Teacher's Edition* for more information.

When you buy a product, you usually consider factors such as price, brand name, quality, and quantity. But do you consider the amount of resources the package uses? Many products come in packages of different types and materials. You might buy a larger pack if you use a lot, or a tiny pack if you like the convenience of individual servings. But how much cardboard, plastic, or glass are you using—or wasting—depending on your choice? How about the trees, petroleum, and other resources needed to make those packages? In this lab, you will compare three sets of packages that hold the same amount of juice to determine how your decisions about packaging affect the use of resources.

Problem Which packaging conserves resources the best?

Materials
- 1 1.89-L (64-fl. oz) cardboard juice carton
- 1 946-mL (32-fl. oz) cardboard juice carton
- 1 240-mL (8-fl. oz) cardboard juice carton
- scissors
- metric ruler

CAUTION: *Be careful when using scissors.*

Skills Observing, Measuring, Calculating, Comparing and Contrasting, Relating Cause and Effect, Drawing Conclusions

Procedure

Part A: Determine the Amount of Material in Each Package

1. Work in groups of three or four. Use scissors to cut apart the three cartons your teacher gives your group. Then spread each one out as you see in Figure 1.

2. Measure the dimensions of the cartons with the ruler.

3. Calculate the area of each carton. Use these equations:
   - Area of a rectangle:
     $A = l \times w$
     (l = length; w = width)
   - Area of a square:
     $A = s^2$
     (s = length of a side of the square)

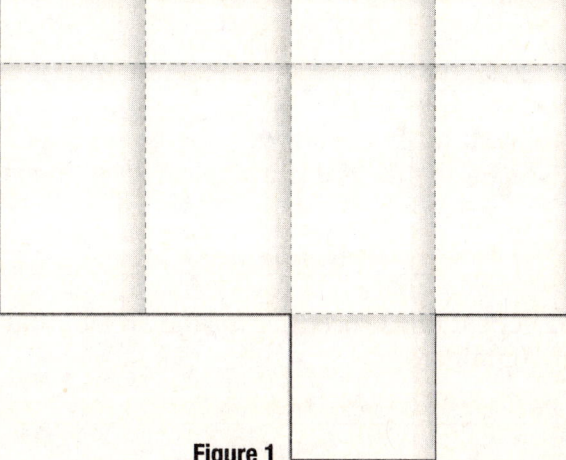

Figure 1

Students' calculations will vary. Sample: 19 × 9 = 171; 171 × 4 = 684; 9 × 9 = 81;
81 × 2 = 162; 684 + 162 = 846 sq cm

Name _____ Class _____ Date _____

4. Record the data you calculated in the Data Table.

DATA TABLE Students' data will vary.

	Amount of Cardboard in One Carton	Number of Cartons Needed to Hold 1.89 L	Amount of Cardboard Needed to Hold 1.89 L
1.89 L		1	
946 mL		2	
240 mL		8	

Part B: Compare the Area of Material in the Packages

5. Calculate how much more cardboard is used when you buy 1.89 L of juice in the two 946-mL cartons instead of one 1.89-L carton.
Students' calculations will vary.

Use this procedure:

a.) Subtract the area of material in the 1.89-L carton from the area of material in the two 946-mL cartons.

Students' calculations will vary. Sample: 356 − 846 = 510 sq cm

_____

b.) Divide the answer you get in Part A by the area of material in the 1.89-L carton.

Students' calculations will vary. Sample: 510 ÷ 846 = .60

_____

c.) Multiply the answer you get in Part B by 100. This is how much more material is in the two containers, expressed as a percentage.

Students' calculations will vary. Sample: 60%

6. Repeat this procedure for the area of material in eight small containers.

Students' calculations will vary.

_____
_____
_____
_____
_____

Name _____ Class _____ Date _____

Analyze and Conclude

1. **Comparing and Contrasting** Based on your data, does buying the juice in one large carton or in an eight-pack of small individual cartons use more cardboard? How does buying the juice in two medium-sized cartons compare?

   Buying an eight-pack of small cartons uses more cardboard. Buying two medium-sized cartons uses more cardboard than one large-sized carton but less than an eight-pack.

2. **Relating Cause and Effect** How does buying the juice in several cartons instead of one large carton impact the use of resources?

   More trees will have to be cut down to supply the materials that make up the additional amount of cardboard.

3. **Drawing Conclusions** Suppose you have determined which set of cardboard cartons uses the least resources. Then you find out that the same size carton of juice comes in plastic and glass as well as cardboard. How would you decide which of these containers would be the best choice, in terms of saving resources?

   Making plastic uses petroleum, a nonrenewable resource, but the plastic container may be made of recycled plastic. Likewise, a glass container may be made of recycled glass. Without knowing the source of the container material, it would be difficult to reach an informed decision. The container that can be recycled would be the best choice because recycling decreases the use of resources.

Earth Science Lab Manual ▪ 173

Name _____ Class _____ Date _____

Chapter 5   Weathering, Soil, and Mass Movements

Exploration Lab

Effect of Temperature on Chemical Weathering

See the *Earth Science Teacher's Edition* for more information.

Water is the most important agent of chemical weathering. One way water promotes chemical weathering is by dissolving the minerals in rocks. In this lab, you will examine the effect of temperature on chemical weathering by measuring the rate at which antacid tablets dissolve in water at different temperatures. These tablets contain calcium carbonate, the mineral found in rocks such as limestone and marble.

Problem How does temperature affect the rate of chemical weathering?

Materials
- 250-mL beaker
- thermometer
- hot water (40–50°C)
- ice
- 5 antacid tablets
- stopwatch

Skills Measuring, Using Tables and Graphs, Drawing Conclusions, Inferring

Procedure

1. Use the Data Table to record your measurements.
2. Add a mixture of hot water and ice to the beaker. Use the thermometer to measure the temperature of the mixture. Add either more hot water or more ice until the temperature is between 0°C and 10°C. The total volume of the mixture should be about 200 mL.
3. When the temperature is within the correct range, remove any remaining ice from the beaker. Record the starting temperature of the water in the Data Table. Remove the thermometer from the beaker.
4. Drop an antacid tablet into the beaker. Start the stopwatch as soon as the tablet enters the water. Stop the stopwatch when the tablet has completely dissolved and no traces of the tablet are visible. (Don't wait for the bubbling to stop.) Record the time in the Data Table.
5. Place the thermometer in the beaker and wait for the temperature of the water to stabilize. Record the final temperature of the water in the Data Table.
6. Calculate the average temperature by adding the starting and final temperatures and dividing by 2. Record the result in the Data Table.
7. Repeat Steps 2–6 four more times, once at each of the following temperature ranges: 10–20°C, 20–30°C, 30–40°C, and 40–50°C. Adjust the relative amounts of hot water and ice to produce the correct water temperatures. The total volume of water and ice should always be about 200 mL.

Earth Science Lab Manual

Name _____ Class _____ Date _____

8. On the following graph, add labels for temperature ranges and time intervals. Then plot your data on the graph. Draw a smooth curve through the data points.

DATA TABLE Students' data will vary.

Starting Temperature (°C)	Dissolving Time (s)	Final Temperature (°C)	Average Temperature (°C)

[Graph: Dissolving Time (s) vs. Average Temperature (°C)]

Analyze and Conclude

1. **Analyzing Data** At which temperature did the antacid tablet dissolve most rapidly?

   the highest temperature

2. **Analyzing Data** At which temperature did the antacid tablet dissolve most slowly?

   the lowest temperature

3. **Drawing Conclusions** What is the relationship between temperature and the rate at which antacid tablets dissolve in water?

   Antacid tablets react more rapidly in water as the temperature increases.

4. **Formulating Hypotheses** Based on your observations, form a hypothesis about the relationship between temperature and the rate of chemical weathering.

   As temperature increases, the rate of chemical weathering increases.

Earth Science Lab Manual ▪ 176

Name _____ Class _____ Date _____

5. Designing Experiments How could you test your hypothesis?

Sample answer: Perform a similar experiment using rock that is sensitive to chemical weathering, such as limestone. Test the water for the presence of calcium or carbonate ions.

6. Predicting What would your results have been if you had ground each tablet into a fine powder before dropping it into the water? Would your conclusion be the same or different? Explain.

The dissolving times would have been shorter because grinding increases the surface area exposed to water. The conclusion would be the same because increasing surface area would speed the rate of reaction at each temperature.

7. Inferring Would a limestone building weather more rapidly in Homer, Alaska, or in Honolulu, Hawaii? (Both cities receive about the same amount of precipitation in an average year.) Explain your reasoning.

Honolulu; Hawaii is much closer to the equator than Alaska is, so the average temperature in Honolulu is greater than that in Homer. The higher temperature promotes faster chemical weathering.

Name _____ Class _____ Date _____

Chapter 6 Running Water and Groundwater   Exploration Lab

Investigating the Permeability of Soils

See the *Earth Science Teacher's Edition* for more information.

The permeability of soils affects the way groundwater moves—or if it moves at all. Some soils are highly permeable, while others are not. In this lab, you will determine the permeability of various soils, and draw conclusions about their effect on the movement of water underground.

Problem How does the permeability of soil affect its ability to move water?

Materials
- 100-mL graduated cylinder
- beaker
- small funnel
- 3 pieces of cotton
- samples of coarse sand, fine sand, and soil
- clock or watch with a second hand

Skills Observing, Measuring, Comparing and Contrasting, Analyzing Data, Interpreting Data

Procedure

1. Place a small, clean piece of cotton in the neck of the funnel. Fill the funnel above the cotton with coarse sand. Fill the funnel about two-thirds of the way.

2. Pour water into the graduated cylinder until it reaches the 50-mL mark.

3. With the bottom of the funnel over the beaker, pour the water from the graduated cylinder slowly into the sand in the funnel.

4. In the Data Table, keep track of the time from the second you start to pour the water into the funnel. Measure the amount of time that it takes the water to drain through the funnel filled with coarse sand.

5. Record the time it takes for the water to drain through the sand, in the Data Table.

6. Empty and clean the measuring cylinder, funnel, and beaker.

7. Repeat Steps 1 through 7, first using fine sand, and then using soil.

DATA TABLE Students' data will vary. Students should discover that the coarse sand has the greatest permeability; fine sand has the least.

	Time Needed for Water to Drain Through Funnel	Water Collected in Beaker (mL)
Coarse sand		
Fine sand		
Soil		

Earth Science Lab Manual

Name _____ Class _____ Date _____

Analyze and Conclude

1. **Comparing and Contrasting** Of the three materials you tested, which has the greatest permeability? Which had the least permeability?

   The coarse sand has the greatest permeability. The fine sand has the least permeability.

2. **Analyzing Data** Why were different amounts of water recovered in the beaker for each material tested?

   Some of the water was left behind in the sample. For example, the soil sample trapped water in the organic material and finer grained sediment.

3. **Interpreting Data** What effect would the differences you observed in this lab have on the movement of groundwater through different soils?

   In general, the soil with coarser grain would allow water to move faster.

Name _____ Class _____ Date _____

Chapter 7   Glaciers, Deserts, and Wind   Exploration Lab

Interpreting a Glacial Landscape

See the *Earth Science Teacher's Edition* for more information.

Topographic maps are valuable tools geologists use to interpret landscapes. Especially in the field—when your view can be limited—these maps not only help you determine your location, they can offer a bigger landscape picture than what is actually visible. See how well you can identify glacial features on the map and interpret them to reconstruct geologic history.

Problem How can a topographic map allow you to interpret a glacially formed landscape?

Materials
- Resource 23 in the DataBank
- piece of blank paper

Skills Graphing, Inferring, Drawing Conclusions

Procedure

1. Following line A on the map, sketch a topographic profile of the Lake Fork Valley onto the Topographic Profile sheet on the next page. Place the straight edge of your blank paper along the line and mark in pencil where it meets every fifth contour line (the darker guide contours). Be sure to write the elevation of every fifth contour line along the *y*-axis of the Topographic Profile sheet.
   Students' profiles should show a U-shaped valley.

2. How can you tell from your profile that the valley was formed by a glacier?
   because it is U-shaped

3. Was the valley shaped by a continental ice sheet or by a valley glacier? Explain how you know.
   valley glacier because it occurs in an alpine region with classic valley glacier features

4. Use the map to help you describe the direction the glacier flowed through this valley. How can you tell?
   west to east because the direction is downhill

5. Which letter arrow points to cirques? You can refer to Figure 7 in your textbook for help.
   D

6. The lakes inside cirques are called tarns. Identify the tarns inside the cirques you just found.
   Lonesome Lake, St. Kevin Lake

7. Which letter arrows point to hanging valleys?
   B

Name _____ Class _____ Date _____

8. Which letter arrows point to arêtes?

   C

9. Name a peak on the map that is a horn.

   Galena Mountain

10. Feature E on the map is composed of glacial till. What type of glacial feature is E, and how did it form?

    It is an end moraine made of debris deposited by the stationary face of the valley glacier. Water backed up behind the end moraine as the glacier retreated.

11. Explain how Turquoise Lake formed.

    A glacier carved out a U-shaped valley that ran west-east between the Galena Mountain region to the north and the Sugar Loaf and Bald Eagle Mountain region to the south. The glacier moved down slope, accumulating mass where topography leveled off, carving a deep enough basin for Turquoise Lake to form from melted ice and runoff.

SOUTH   NORTH

Sugar Loaf Mountain   Bear Lake

Topographic Profile sheet

Chapter 8 Earthquakes and Earth's Interior See the *Earth Science Teacher's Edition* for more information. Exploration Lab

Locating an Earthquake

The focus of an earthquake is the actual place within Earth where the earthquake originates. When locating an earthquake on a map, scientists plot the epicenter, the point on Earth's surface directly above the focus. To locate an epicenter, records from three different seismographs are needed.

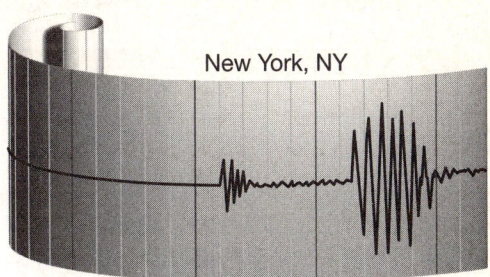

9:00 UTC (Time marks in minutes)

Problem How can you determine the location of an earthquake's epicenter?

Materials
- drawing compass
- world map or atlas

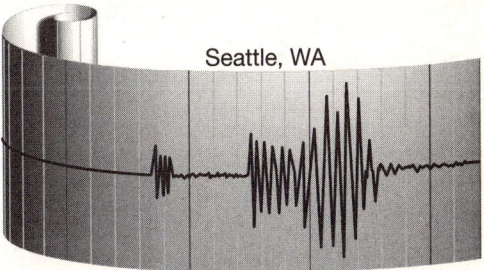

9:00 UTC

Skills Measuring, Interpreting Maps, Interpreting Graphs

Procedure

1. The seismograms shown in Figure 1 recorded the same earthquake. Use the Travel-Time Graph to determine the distance of each station from the epicenter. Record your answers in the Data Table.

2. Refer to a world map or atlas for the locations of the three seismic stations. Place a small dot showing the location of each of the three stations on the map in Figure 2. Neatly label each city on the map.

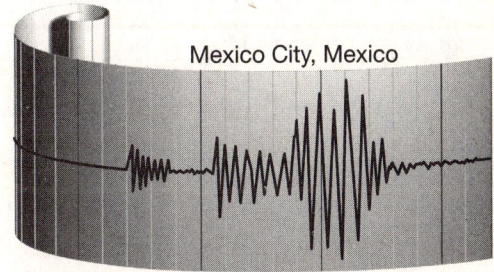

9:00 UTC

Figure 1

3. On the map in Figure 2, use a drawing compass to draw a circle around each of the three stations. The radius of the circle, in miles, should be equal to each station's distance from the epicenter. Use the scale on the map to set the distance on the drawing compass for each station. **CAUTION:** *Use care when handling the drawing compass.*

DATA TABLE

	New York	Seattle	Mexico City
Elapsed time between first P and first S waves	5.5	4.0	3.5
Distance from epicenter in miles	2400	1400	1200

Earth Science Lab Manual ▪ 183

Name _____ Class _____ Date _____

Analyze and Conclude

1. **Using Graphs** How far from the epicenter are the three cities located?

   New York—approximately 2400 miles;

   Seattle—approximately 1400 miles;

   Mexico City—approximately 1200 miles

2. **Calculating** What would the distances from the epicenter to the cities be in kilometers?

   New York—approximately 3864 km;

   Seattle—approximately 2254 km;

   Mexico City—approximately 1932 km

3. **Interpreting Maps** What is the approximate latitude and longitude of the epicenter of the earthquake that was recorded by the three stations? Use the map in Figure 3.

   28° N latitude and 112°30′W longitude

4. **Drawing Conclusions** On the New York seismogram, the first P wave was recorded at 9:01 UTC. UTC is the international standard on which most countries base their time. At what time (UTC) did the earthquake actually occur? Explain.

   8:54 UTC; There is a 7-minute difference between the actual time of the

   earthquake and when it was recorded in New York.

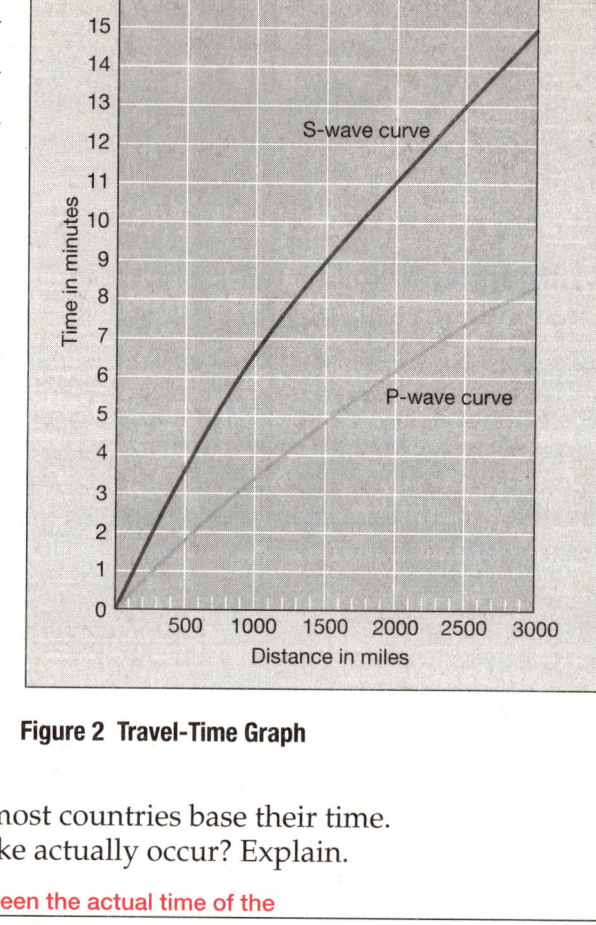

Figure 2 Travel-Time Graph

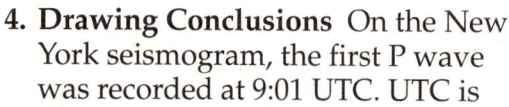

Figure 3

Chapter 9   Plate Tectonics   See the *Earth Science Teacher's Edition* for more information.

Exploration Lab

Paleomagnetism and the Ocean Floor

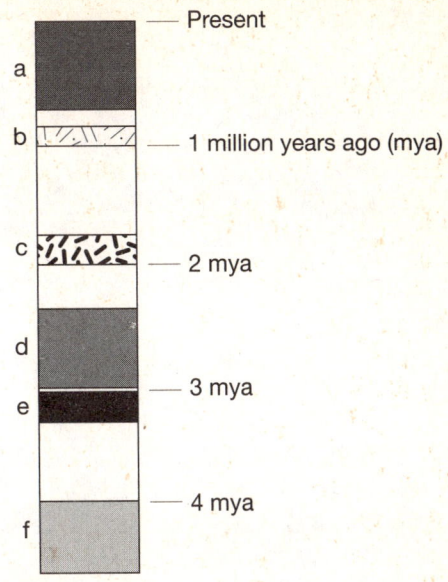

Figure 1

In the continental drift hypothesis, the ocean floors were not really involved. The hypothesis proposed that the continents moved through the oceans like icebreaking ships plowing through ice. Later studies of the oceans provided one of the keys to the plate tectonic theory. You will observe how the magnetic rocks on the ocean floor can be used to understand plate tectonics.

Problem   How are the paleomagnetic patterns on the ocean floor used to determine the rate of seafloor spreading?

Materials
- metric ruler
- calculator

Skills   Measuring, Interpreting Diagrams, Calculating

Procedure

1. Scientists have reconstructed Earth's magnetic polarity reversals over the past several million years. A record of these reversals is shown in Figure 1. Periods of normal polarity, when a compass would have pointed north as it does today, are shown in grayscale. Periods of reverse polarity are shown in white. Record the number of times Earth's magnetic field has had reversed polarity in the last 4 million years.

    five times

2. The three diagrams in Figure 2 on the next page illustrate the magnetic polarity reversals across sections of the mid-ocean ridges in the Pacific, South Atlantic, and North Atlantic oceans. Periods of normal polarity are shown in the same grayscale in the illustration above. Observe that the patterns of polarity in the rock match on either side of the ridge for each ocean basin.

3. On the three ocean-floor diagrams, identify and mark the periods of normal polarity with the letters *a–f*. Begin at the ridge crest and label along both sides of each ridge. (*Hint:* The left side of the South Atlantic has already been done and can act as a guide.)

4. Using the South Atlantic as an example, label the beginning of the normal polarity period c, "2 million years ago," on the left sides of the Pacific and North Atlantic diagrams.

Earth Science Lab Manual   ▪   **185**

Name _____ Class _____ Date _____

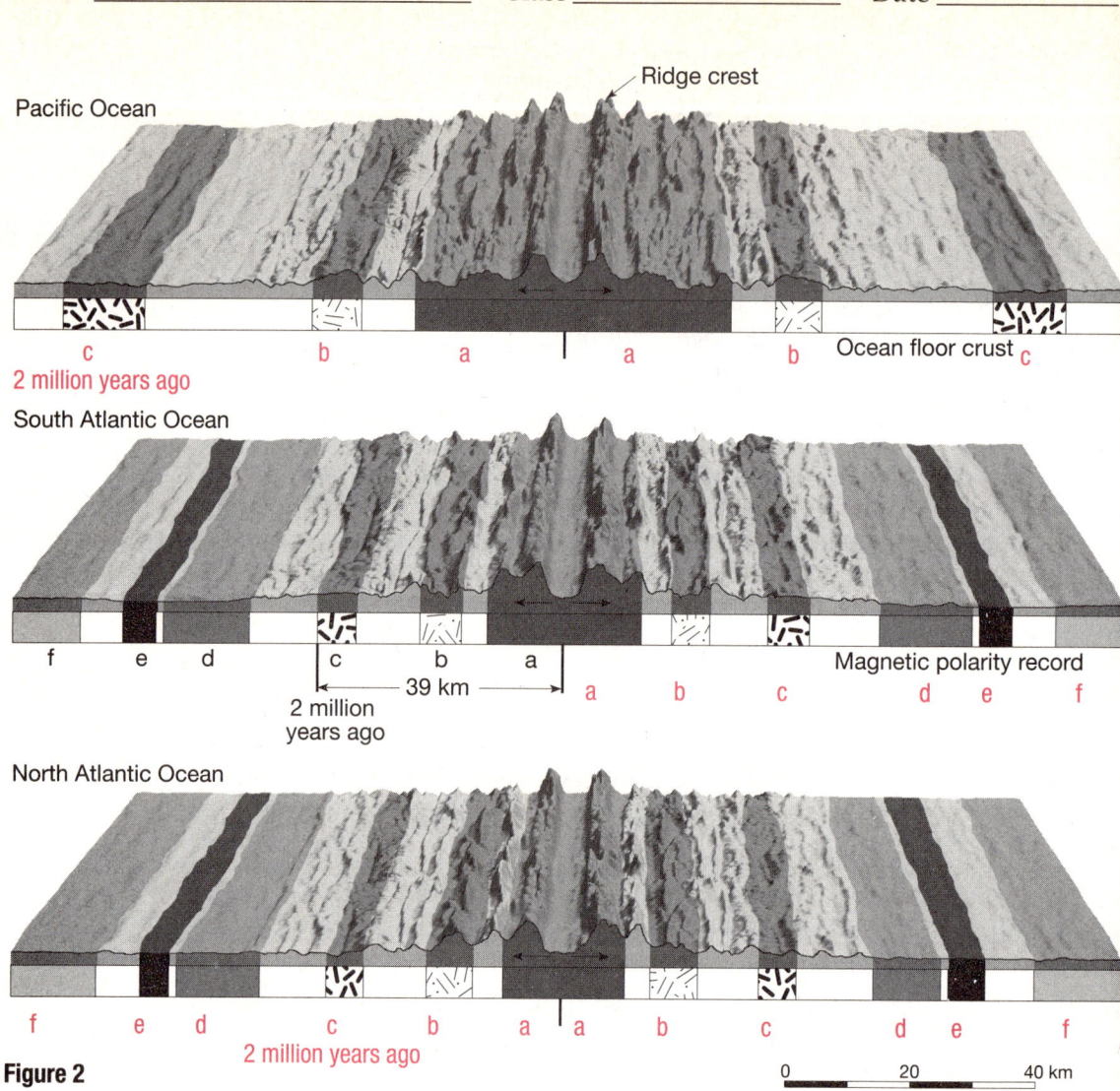

Figure 2

5. Using the distance scale shown with the ocean floor diagrams, determine which ocean basin has spread the greatest distance during the last 2 million years.

   The Pacific Ocean basin has spread the greatest distance.

6. Refer to the distance scale. Notice that the left side of the South Atlantic basin has spread approximately 39 kilometers from the center of the ridge crest in 2 million years.

Analyze and Conclude

1. **Analyzing Data** How many kilometers has the left side of the Pacific basin spread in 2 million years?

   approximately 80 km

Name _____ Class _____ Date _____

2. Analyzing Data How many kilometers has the left side of the North Atlantic basin spread in 2 million years?

approximately 37 km

3. Inferring How many kilometers has each ocean basin opened in the past 2 million years?

Pacific: approximately 160 km; North Atlantic: approximately 74 km;

South Atlantic: approximately 78 km

4. Calculating If both the distance that each ocean basin has opened and the time it took to open that distance are known, the rate of seafloor spreading can be calculated. Determine the rate of seafloor spreading for the South Atlantic Ocean basin in centimeters per year. (*Hint:* To determine the rate of spreading in centimeters per year for each ocean basin, first convert the distance from kilometers to centimeters and then divide this distance by the time, 2 million years.)

The spreading rate for the South Atlantic Ocean basin is 3.9 cm/yr.

5. Calculating Determine the rate of seafloor spreading for the North Atlantic and Pacific Ocean basins.

The spreading rate for the Pacific Ocean basin is 8.0 cm/yr. The spreading rate for the North

Atlantic Ocean basin is 3.7 cm/yr.

6. Drawing Conclusions Which ocean basin is spreading the fastest? The slowest?

Fastest: Pacific; slowest: North Atlantic

7. Inferring Do ocean basins spread uniformly over the entire basin? Explain.

No; large basins such as the Pacific Ocean basin are spreading at various rates along

their spreading ridges.

Earth Science Lab Manual

Chapter 10   Volcanoes and Other Igneous Activity

Exploration Lab

Melting Temperatures of Rocks

See the *Earth Science Teacher's Edition* for more information.

Measurements of temperatures in wells and mines have shown that Earth's internal temperatures increase with depth. Recall that this rate of temperature increase is called the geothermal gradient. Although the geothermal gradient varies from place to place, it is possible to calculate an average. In this lab, you will investigate Earth's internal temperatures and the temperatures at which rocks melt. You will also investigate the effect of water on the melting temperatures of rock.

Problem How can rocks melt to form magma in the crust and uppermost mantle?

Materials
- colored pencils (three different colors)
- ruler

Skills Analyzing Data, Graphing, Calculating

Procedure

1. Use the Temperature Curves graph on the next page to plot the average temperature gradient for Earth's interior.

2. Plot the temperature values from Data Table 1 on the graph. Then draw a single best-fit line through the points with a colored pencil. Extend your line from the surface to 200 kilometers. Label the line "Temperature Gradient."

3. The melting temperature of a rock changes as pressure increases deeper within Earth. The approximate melting points of the igneous rocks granite and basalt under various pressures (depths) have been determined in the laboratory and are shown in Data Table 2. Granite and basalt were used because they are common materials in the upper layer of Earth. Plot the melting temperatures from Data Table 2 on the same graph. Use a different colored pencil to plot each set of points and draw the best-fit lines.

Data Table 1 Idealized Internal Temperatures of Earth

Depth (kilometers)	Temperature (°C)
0	20
25	600
50	1000
75	1250
100	1400
150	1700
200	1800

Data Table 2 Melting Temperatures of Granite (with water) and Basalt at Various Depths Within Earth

Granite (with water)		Basalt	
Depth (km)	Melting Temperature (°C)	Depth (km)	Melting Temperature (°C)
0	950	0	1100
5	700	25	1160
10	660	50	1250
20	625	100	1400
40	600	150	1600

4. Label the two lines "Melting Curve for Wet Granite" and "Melting Curve for Basalt."

Analyze and Conclude

1. **Using Graphs** Does the rate of increase of Earth's internal temperature stay the same or change with increasing depth?
   change with increasing depth

2. **Using Graphs** Is the rate of temperature increase greater from the surface to 100 km or below 100 km?
   greater from the surface to 100 km

3. **Interpreting Data** What is the temperature at 100 km below the surface?
   1400°C

4. **Calculating** Use the data and graph to calculate the average temperature gradient for the upper 100 km of Earth in °C/100 km and in °C/km.
   1400°C/100 km; 14°C/km

5. **Drawing Conclusions** Based on your data, at approximately what depth within Earth would wet granite reach its melting temperature and begin to form magma? Explain.
   Approximately 25 km; at this depth, the internal temperature of Earth is approximately the same as the melting temperature for wet granite. (The two curves intersect.)

6. **Drawing Conclusions** Based on your data, at what depth will basalt have reached its melting temperature and begin to form magma?
   Basalt will reach its melting temperature at approximately 100 km in depth.

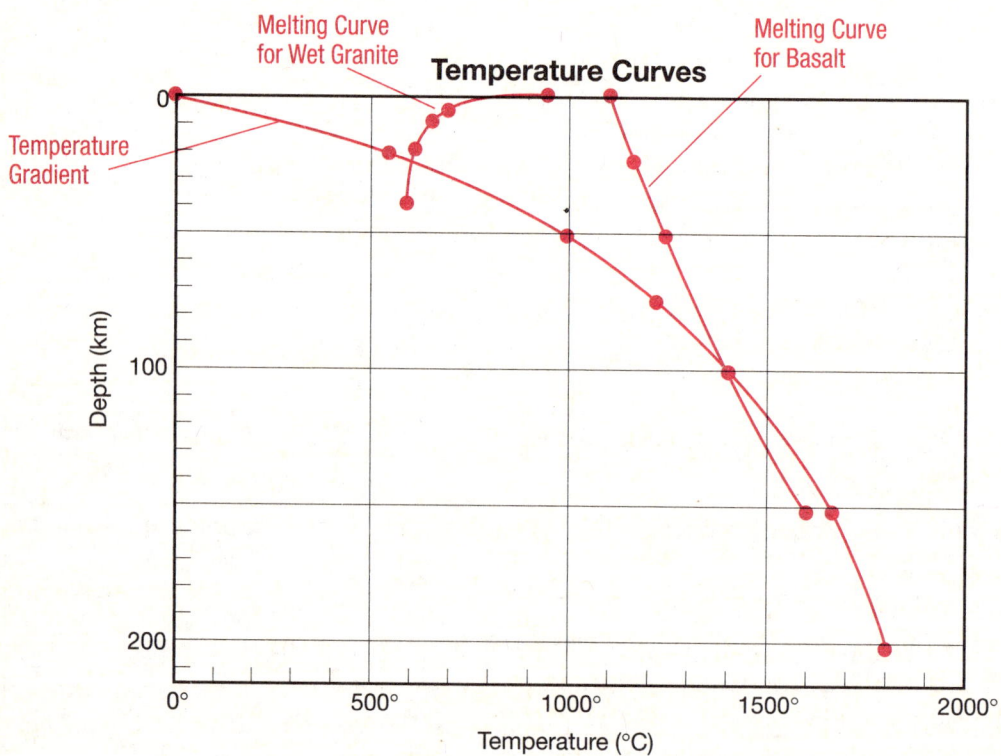

Earth Science Lab Manual ▪ 190

Name _____ Class _____ Date _____

Chapter 11   Mountain Building   See the *Earth Science Teacher's Edition* for more information.   **Exploration Lab**

Investigating Anticlines and Synclines

The axial plane of a fold is an imaginary plane drawn through the long axis of a fold. The axial plane divides the fold into two halves called limbs as shown in Figure 1. In a symmetrical fold, the limbs are mirror images of each other and move away at the same angle. In an asymmetrical fold, the limbs dip or tilt at different angles. Folds do not continue forever. Where folds "die out" and end, the axis is no longer horizontal, and the fold is said to be plunging, as shown in Figure 2. A geologic principle known as the principle of superposition states that in most situations with layered rocks, the oldest rocks are at the bottom of the sequence.

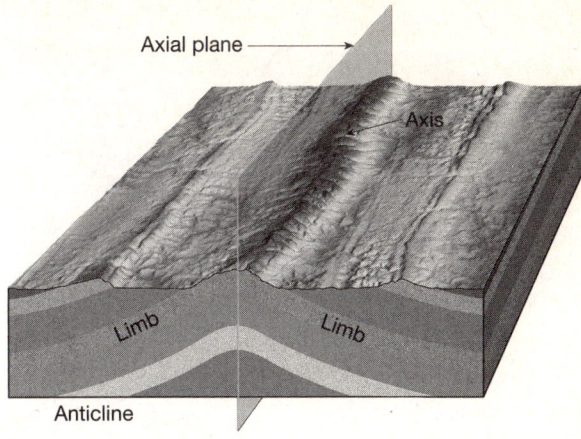

Figure 1 Horizontal Axis

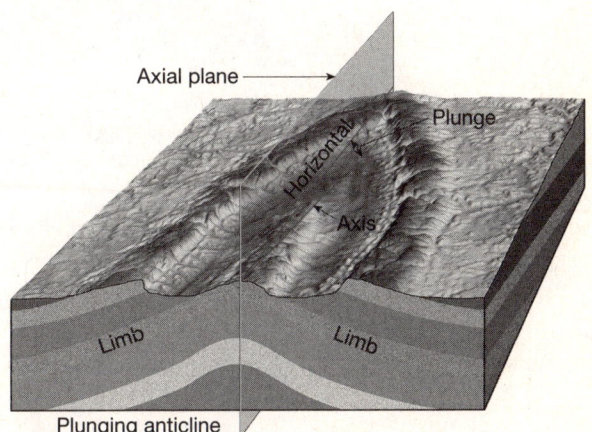

Figure 2 Plunging Axis

Problem How are rocks oriented in anticlines and synclines?

Materials
- protractor

Skills Observing, Measuring, Classifying, Interpreting Diagrams

Procedure

1. Study the two diagrams, labeled Fold A and Fold B in Figures 3 and 4.

2. Use a protractor to measure the angles of the rock layers in both limbs of Fold A. Repeat your measurements for both limbs of Fold B. For consistency, measure the angles on both folds at the surface between layers 3 and 4. Record the measurements of the angles.

3. Use Figures 3, 4, and 5 to determine what types of folds are shown by Fold A and Fold B.

4. Anticlines and synclines are linear features caused by compressional stresses. Two other types of folds—domes and basins—are often nearly circular and result from vertical displacement. Uplift produces domes like those shown in Figure 3. A basin is a downwarped structure, as shown in Figure 4.

5. Complete the three sides of the blank block diagram on the right to show an eroded fold consistent with the rock layer shown on the right side of the block.

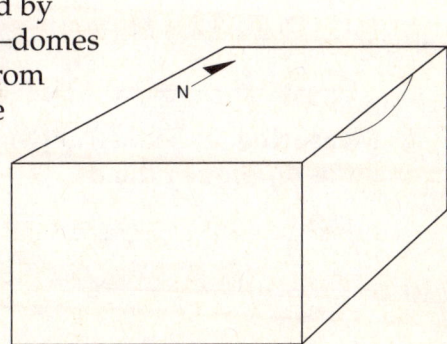

Earth Science Lab Manual • 191

Name _____ Class _____ Date _____

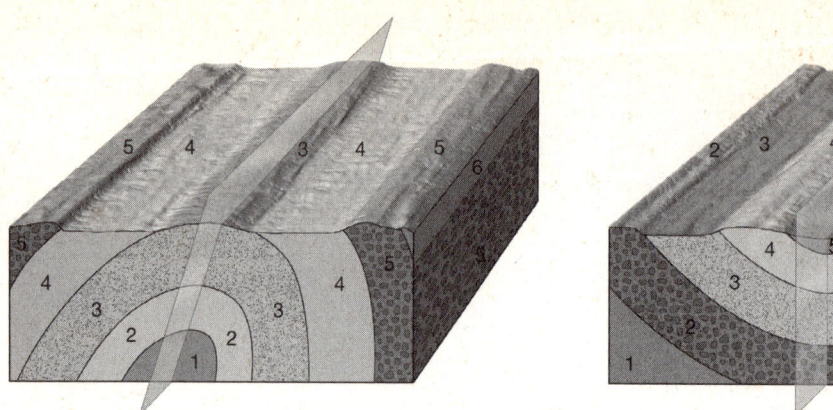

Figure 3 Fold A

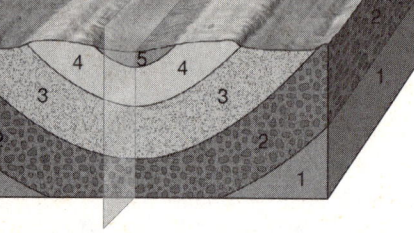

Figure 4 Fold B

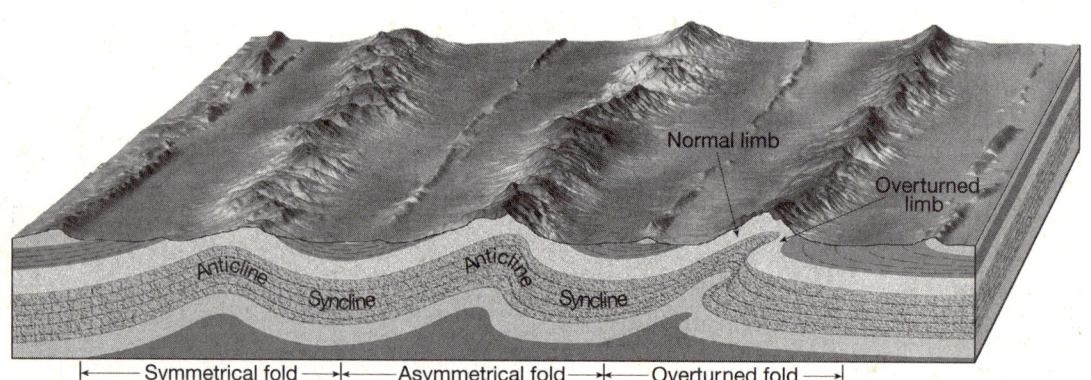

Figure 5 Anticlines and Synclines

Analyze and Conclude

1. **Interpreting Diagrams** What type of fold is shown by Fold A? In what direction do the limbs dip or tilt from the axial plane?

   an anticline; away from the axial plane

2. **Interpreting Diagrams** What type of fold is shown by Fold B? In what direction do the limbs dip or tilt from the axial plane?

   a syncline; towards the axial plane

3. **Drawing Conclusions** In Fold A, which rock layer is the oldest shown? Which rock layer is the youngest shown?

   In Fold A, the oldest rock layer shown is 1; the youngest is 6.

4. **Measuring** In Fold A, at what angle are the rock layers in both limbs dipping or tilted?

   In Fold A the left limb dips at about 50°, and the right limb dips about 80°.

Name _____ Class _____ Date _____

5. **Drawing Conclusions** In Fold B, which rock layer is the oldest shown? Which rock layer is the youngest shown?

   In Fold B, the oldest rock layer shown is 1; the youngest is 5.

6. **Measuring** In Fold B, at what angle are the rock layers in both limbs dipping or tilted?

   In Fold B, both limbs dip at about 45°.

7. **Classifying** What type of fold did you draw in the blank block diagram?

   Fold is similar to Fold B, a syncline with horizontal axis.

8. **Observing** Is Fold A symmetrical or asymmetrical? Is Fold B symmetrical or asymmetrical?

   Fold A is asymmetrical; Fold B is symmetrical.

9. **Observing** Is Fold A plunging or nonplunging? Is Fold B plunging or nonplunging?

   Both Fold A and Fold B are nonplunging folds.

10. **Applying Concepts** If you walk away from the axis on an eroded anticline, do the rocks get older or younger? How do the ages of the rocks change as you walk away from the axis in a syncline?

    On an eroded anticline, the rock layers get younger as you walk away from the axial plane. On an eroded syncline, the rock layers get older as you walk away from the axial plane.

Earth Science Lab Manual

Name _____ Class _____ Date _____

Chapter 12 Geologic Time   See the *Earth Science Teacher's Edition* for more information.   Exploration Lab

Fossil Occurrence and the Age of Rocks

Groups of fossil organisms occur throughout the geologic record for specific intervals of time. This time interval is called the fossil's range. Knowing the range of the fossils of specific organisms or groups of organisms can be used to relatively date rocks and sequences of rocks. In this laboratory exercise, you will use such information to assign a date to a hypothetical unit of rock.

Problem How can the occurrence of fossils and their known age ranges be used to date rocks?

Materials
- Resource 10 in the DataBank

Skills Interpreting Diagrams, Graphing, Hypothesizing, Inferring

Procedure
1. A section of rock made up of layers of limestone and shale has been studied and samples have been taken. A large variety of fossils were collected from the rock samples. Make a bar graph using the information shown in the Data Table. Begin by listing the numbers of the individual fossils on the *x*-axis. Use the Geologic Time Scale in the DataBank to list the time units of the Geologic Time Scale on the *y*-axis.

2. Transfer the range data of each fossil onto the graph. Draw an X in each box, beginning at the oldest occurrence of the organism up to the youngest occurrence. Shade in the marked boxes. You will end up with bars depicting the geologic ranges of each of the fossils listed.

3. Examine the graph. Are there any time units that contain all of the fossils listed? Write this time period at the bottom of the graph. Devonian

DATA TABLE

	Type of Fossil	Oldest Occurrence	Youngest Occurrence
1	Foraminifera	Silurian	Quaternary
2	Bryozoan	Silurian	Permian
3	Gastropod	Devonian	Pennsylvanian
4	Brachiopod	Silurian	Mississippian
5	Bivalve	Silurian	Permian
6	Gastropod	Ordovician	Devonian
7	Trilobite	Silurian	Devonian
8	Ostracod	Devonian	Tertiary
9	Brachiopod	Cambrian	Devonian

Earth Science Lab Manual ■ 195

Name _____ Class _____ Date _____

	1	2	3	4	5	6	7	8	9
Neogene/Quaternary	X								
Paleogene	X							X	
Cretaceous	X							X	
Jurassic	X							X	
Triassic	X							X	
Permian	X	X			X			X	
Pennsylvanian	X	X	X		X			X	
Mississippian	X	X	X	X	X			X	
Devonian	X	X	X	X	X	X	X	X	X
Silurian	X	X		X	X	X	X		X
Ordovician						X			X
Cambrian									X

Analyze and Conclude

1. **Reading Graphs** What is the age of the hypothetical rock layer that these fossils were collected from?

   The rock is from the Devonian period and is 354 to 417 million years old.

2. **Inferring** Based on the age determined, do you think that this group of fossils could be considered index fossils? Why or why not?

   No, because the age is Devonian, which covers 31 million years. In terms of geologic time, this is too long to meet the requirements for an index fossil.

3. **Inferring** A species of the trilobite listed in line 7 of the Data Table (*Paciphacops logani*) is limited to rocks of lower Devonian age. Trilobite fossils are widespread throughout North America. Can this fossil be considered an index fossil? Why or why not?

   Yes, because it is widespread, which indicates that it was abundant, and because it lived over a short geologic time span.

4. **Connecting Concepts** These fossils were collected from limestone and shale rocks. Based on what you have learned about the formation of these rock types, what type of environment did these organisms live in?

   The environment was an ocean.

5. **Understanding Concepts** Shale often contains fossils of leaves. If the gastropods listed in line 3 and line 6 were collected from shale containing leaf fossils, could you use radiocarbon dating to assign a numerical date to this rock unit? Explain.

   No, radiocarbon methods can only be used on materials that are less than 75,000 years old. The gastropods are much older than that.

Earth Science Lab Manual ▪ 196

Name _____ Class _____ Date _____

Chapter 13 Earth's History

Application Lab

See the *Earth Science Teacher's Edition* for more information.

Modeling the Geologic Time Scale

Applying the techniques of geologic dating, the history of Earth has been subdivided into several different units that provide a meaningful time frame. The events that make up Earth's history can be arranged within this time frame to provide a clearer picture of the past. The span of a human life is like the blink of an eye compared to the age of Earth. Because of this, it can be difficult to comprehend the magnitude of geologic time.

Problem How can the geologic time scale be represented in a way that allows a clearer visual understanding?

Materials
- strip of adding machine paper measuring 5 meters or longer
- meter stick or metric measuring tape
- Resource 10 in the DataBank

Skills Measuring, Calculating, Interpreting Diagrams

Procedure

1. Obtain a piece of adding machine paper slightly longer than 5 meters in length. Draw a line at one end of the paper and label it "Present."

2. Using the following scale, construct a timeline by completing Steps 3 and 4.

   Scale

   1 meter = 1 billion years

   10 centimeters = 100 million years

   1 centimeter = 10 million years

   1 millimeter = 1 million years

3. Using the Geologic Time Scale as a reference, divide your timeline into the eons and eras of geologic time. Label each division with its name and indicate its absolute age.

4. Using the scale, plot and label the plant and animal events on your timeline that are listed on the Geologic Time Scale.

Analyze and Conclude

1. **Calculating** What fraction or percent of geologic time is represented by the Precambrian eon?

   The Precambrian eon represents 3960/4560, or 87%.

Name _____ Class _____ Date _____

2. **Explaining** Using your text and class notes as references, explain why the approximate time of 540 million years ago was selected to mark the end of the Precambrian eon and the beginning of Phanerozoic eon.

   appearance of the first organisms with shells and other hard parts

3. **Inferring** Suggest one reason why the periods of the Cenozoic era have been further subdivided into several epochs with reasonably reliable accuracy.

   The Cenozoic is the most recent era, so the fossils from this time period are the best preserved. This allows scientists to study the fossils in detail and observe even small changes over time.

4. **Analyzing Data** How many times longer is the whole of geologic time than the time represented by the 5000 years of recorded history?

   4,500,000,000/5000 = 900,000 times longer

5. **Calculating** For what fraction or percent of geologic time have land plants been present on Earth?

   443,000,000/4,500,000,000 = 0.098 × 100 = 9.8% of the time

Name _____ Class _____ Date _____

Chapter 14   The Ocean Floor   Exploration Lab

Modeling Seafloor Depth Transects

See the *Earth Science Teacher's Edition* for more information.

Oceanographers use a number of methods to determine the depth and topography of the ocean floor. Technology—such as sonar, satellites, and submersibles—has allowed scientists to produce detailed maps of the ocean floor in each ocean basin. In this lab, you will model a seafloor depth transect to determine the topography of an ocean basin created by your classmates.

Problem How can the topography of an ocean basin be determined?

Materials
- shoe box
- modeling clay
- aluminum foil
- metric ruler
- scalpel
- Resource 18 in the DataBank

Skills Measuring, Graphing, Inferring, Drawing Conclusions

Procedure

Part A: Making a Model of the Seafloor

1. Examine Figures 1 and 2 and Resource 18 in the DataBank to determine which area of the ocean floor you and your group will model. Be sure to identify the specific features that would be found in the area you choose to model. For example, if you were to model the continental margin you would want to include the continental shelf, continental slope, continental rise, and some submarine canyons in your model. If you were to model the ocean basin floor, you would want to include abyssal plains, trenches, seamounts, and guyots. Do not discuss the plan for your model with students outside your group.

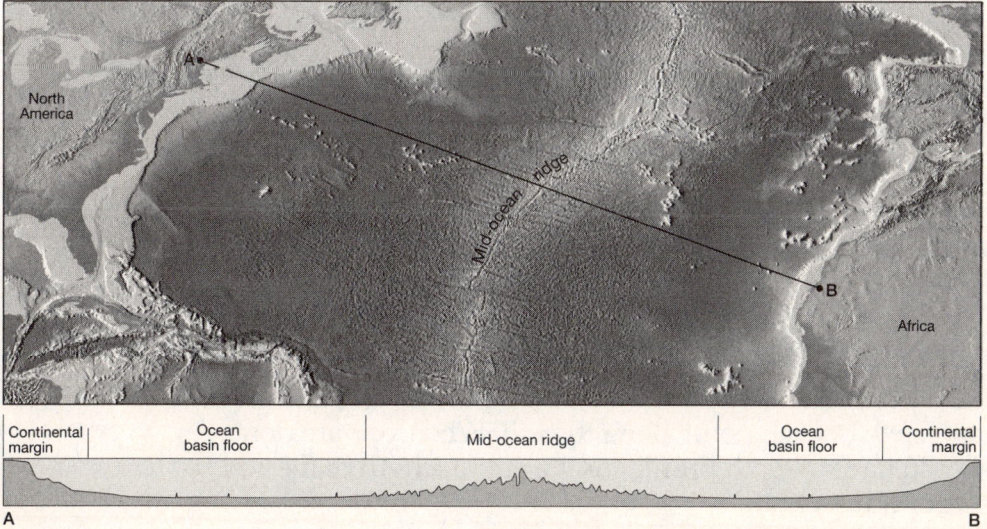

Figure 1

Earth Science Lab Manual ▪ 199

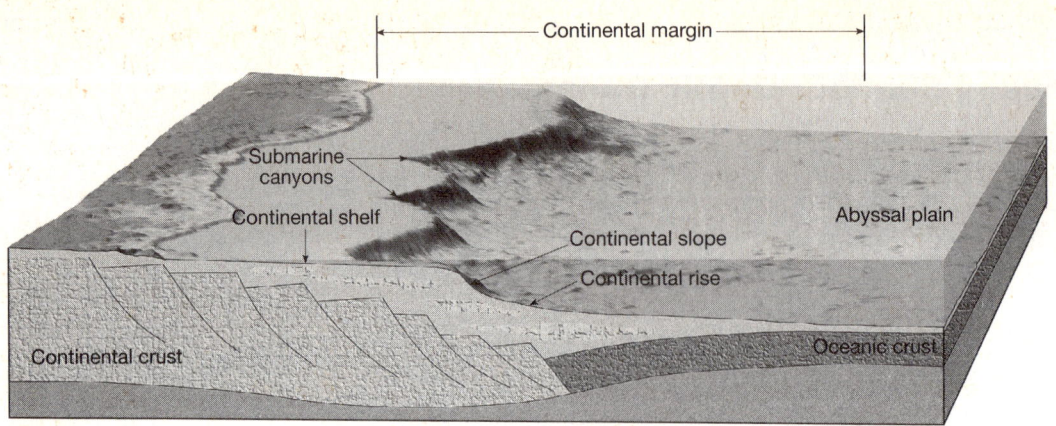

Figure 2

2. Once you have determined which area of the ocean floor you will model, use the clay to make a contoured model of the seafloor inside the shoebox.

3. Cover the box with its top and exchange boxes with another group from your class. Do not remove the top of the box that you receive from another group.

Part B: Completing a Depth Transect

4. Obtain a piece of aluminum foil that is large enough to cover the top of the shoebox and fold over the sides of the box about an inch all the way around.

5. Spread the foil flat on your lab table. Place the ruler lengthwise on the foil, parallel to the edge of the foil. The ruler can be in the middle of the foil or off to the side. The line formed by the edge of the ruler will be your transect line.

6. Use a pencil to make tick marks on the foil every centimeter along the entire length of the foil.

7. Hold the foil in place over the top of the box. Quickly and carefully remove the top of the box and set the foil piece down in place of the top. Do not look in the box. Secure the foil in place on top of the box by turning down the foil over the sides of the box. Be sure the foil is tight across the top.

8. Make tick marks along the x-axis of the graph once every centimeter. Make tick marks along the y-axis every half of a centimeter.

9. Use the scalpel to carefully make a slit in the foil along the first centimeter mark. **CAUTION:** *The scalpel is extremely sharp. Handle it carefully.* After cutting the foil, gently place the ruler through the slit until it makes contact with the clay in the box. Be sure to hold the ruler straight. Take the depth measurement. Record your data on the graph on the next page.

10. Repeat Step 9 for each point along the foil. When you are done, you should have a depth profile for the entire length of the box along your transect line.

11. Remove the foil from the box and examine the contour of the model.

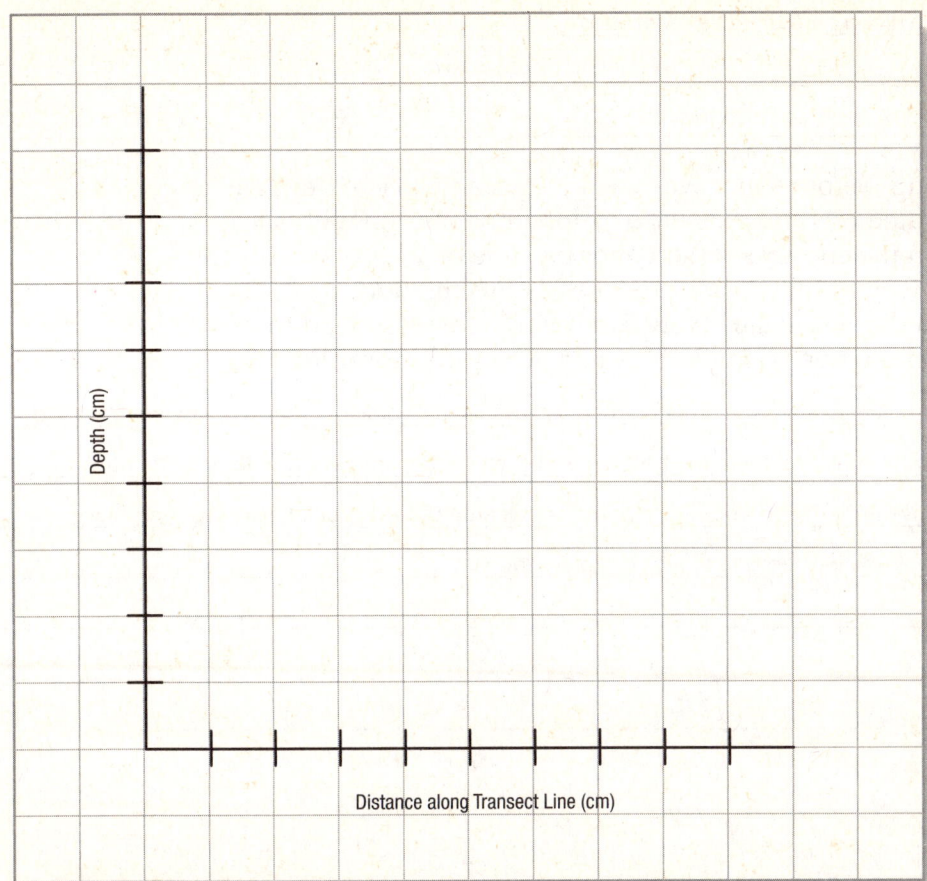

Analyze and Conclude

1. **Inferring** Based on your contour profile, what part of the ocean floor was being modeled? Check your answer with the group that created the model. Were you correct? Why or why not?

   continental margin, ocean basin floor, or mid-ocean ridge; Students should check with the group to see if they were correct. Students should explain if they were correct and why they were or weren't.

2. **Comparing** How does the profile on the graph compare with the contour of the model? Are there any major features in the model that did not appear on your graph? Why or why not?

   Students' responses will depend on which area of the ocean floor was contoured and what area of the model they chose to measure. If major features are missing from the measurements, it might be because only one transect line was measured. The larger the number of transects, the more accurate the results.

3. **Analyzing Data** What could you have done to make your profile match the contour more accurately?

   collect data along additional transect lines

4. **Explaining** Before sonar was used to measure ocean depth, a less sophisticated method was used. A long line of rope with a lead weight on the end was tossed over the side of a ship and lowered until the weight hit the bottom. How is this method similar to what you did in the lab? How can the rope method lead to inaccuracies when trying to build an ocean-floor profile?

   The methods are similar in that no technology was used to collect data. A ruler was extended into the model until it reached the bottom, and then a measurement was taken. The data collected give information about only a very small area of the model. Large topographic features could be missed due to low sampling ratios.

Name _____ Class _____ Date _____

Chapter 15 Ocean Water and Ocean Life **Exploration Lab**

How Does Temperature Affect Water Density?

See the *Earth Science Teacher's Edition* for more information.

Ocean water temperatures vary from equator to pole and change with depth. Temperature, like salinity, affects the density of seawater. However, the density of seawater is more sensitive to temperature fluctuations than salinity. Cool surface water, which has a greater density than warm surface water, forms in the polar regions, sinks, and moves toward the tropics.

Problem How can you determine the effects of temperature on water density?

Materials
- 2 100-mL graduated cylinders
- 2 test tubes
- 2 beakers
- food coloring or dye
- stirrer
- ice
- tap water
- graph paper
- colored pencils

Skills Observing, Graphing, Inferring, Drawing Conclusions

Procedure
Part A

1. In a beaker, mix cold tap water with several ice cubes. Stir until the water and ice are well mixed.

2. Fill the graduated cylinder with 100 mL of the cold water from the beaker. The graduated cylinder should not contain any pieces of ice.

3. Put 2 to 3 drops of food coloring or dye in a test tube and fill it half full with hot tap water.

4. Pour the contents of the test tube slowly into the graduated cylinder. Record your observations.

   Sample observation: The warm water with the dye did not sink below the cold water in the beaker.

5. Add a test tube full of cold tap water to a beaker. Mix in 2 to 3 drops of food coloring dye and a handful of ice to the beaker. Stir the solution thoroughly.

6. Fill the test tube half full of the solution from Step 5. Do not allow any ice into the test tube.

7. Fill the second graduated cylinder with 100 mL of hot tap water.

Earth Science Lab Manual ▪ 203

Name _____ Class _____ Date _____

8. Pour the test tube of cold liquid slowly into the cylinder of hot water. Record your observations.

   Sample observation: The cold water with the dye sank beneath the warm water in the beaker.

   _____

9. Clean the glassware and return it along with other materials to your teacher.

Part B

1. Use the graph to plot surface temperature and density.

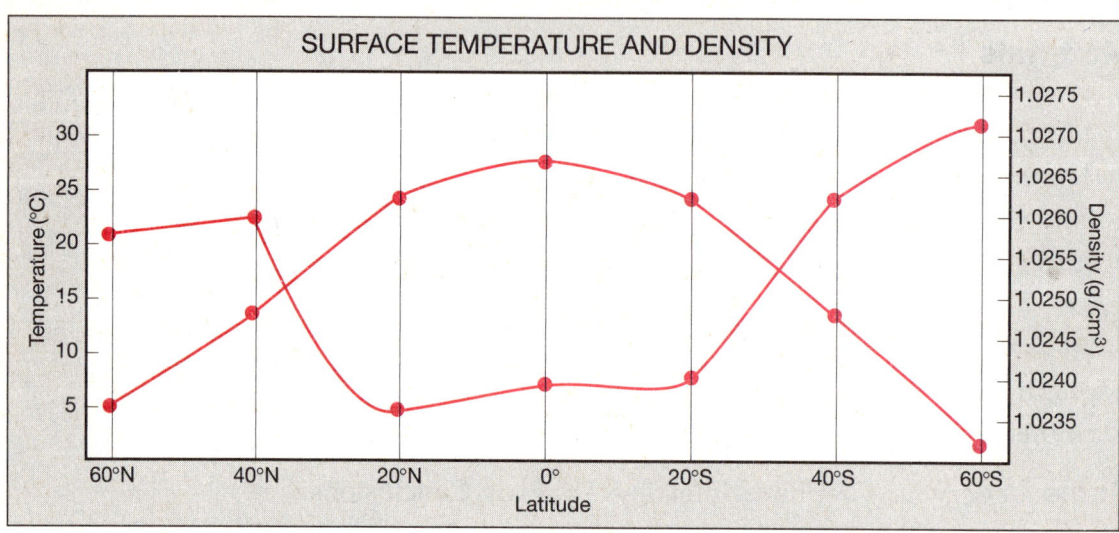

2. Using the information in the Data Table, plot a line on the graph for temperature. Using a different colored pencil, plot a line for density on the same graph.

DATA TABLE Idealized Ocean Surface Water Temperatures and Densities at Various Latitudes

Latitude	Surface Temperature (°C)	Surface Density (g/cm^3)
60°N	5	1.0258
40°N	13	1.0259
20°N	24	1.0237
0°	27	1.0238
20°S	24	1.0241
40°S	15	1.0261
60°S	2	1.0272

Analyze and Conclude

1. **Observing** What differences did you observe in the behavior of the water samples from Steps 4 and 8? Which water sample was the most dense in each experiment?

    The hot water remained on the surface of the cooler water. The cold water sank beneath the warmer water in the cylinder. The cold water has the highest density in both experiments.

2. **Inferring** How does temperature affect the density of water?

    As temperature decreases, water density increases.

3. **Drawing Conclusions** If two water samples of equal mass had equal salinities, which sample would be more dense: Water Sample A, which has a temperature of 25°C, or Water Sample B, which has a temperature of 14°C?

    Water Sample B

4. **Interpreting Diagrams** Describe the density and temperature characteristics of water in equatorial regions. Compare these characteristics to water found in polar regions.

    The water in equatorial regions is warm, and the density is low. The water in polar regions is cold, and the density is high.

5. **Inferring** What is the reason that higher average surface densities are found in the Southern Hemisphere?

    The temperature of surface waters is cooler.

Name _____ Class _____ Date _____

Chapter 16 The Dynamic Ocean

Graphing Tidal Cycles

Exploration Lab
See the *Earth Science Teacher's Edition* for more information.

Tides are the cyclical rise and fall of sea level caused by the gravitational attraction of Earth to the moon and, to a lesser extent, to the sun. Gravitational pull creates a bulge in the ocean on the side of Earth nearest the moon. This inertia creates a similar bulge on the opposite side of Earth from the moon. Tides develop as the rotating Earth moves through these bulges, causing periods of high and low water. In this lab, you will make a graph of tidal data to determine whether an area has diurnal, semidiurnal, or mixed tides.

Problem How can you determine the tidal pattern an area experiences?

Materials
- pencil

Skills Graphing, Interpreting Data, Inferring, Drawing Conclusions

Procedure
1. Use the information in the Data Table to make a graph of the tidal cycle.

Tidal Curve for Long Beach, January 2003

Analyze and Conclude

1. **Applying Concepts** What tidal pattern does this area experience? Explain how you determined this.

   semidiurnal, which can be determined by studying the graph produced from the data; the area has two high tides and two low tides in an approximately 24-hour period, with the high tides being similar height and the low tides being similar height.

2. **Calculating** What is the greatest tidal range for the data you graphed? What is the smallest tidal range? What types of tides correspond to each of these tidal ranges?

   6.4 feet; 2.6 feet; spring tides and neap tides respectively

Earth Science Lab Manual ▪ **207**

Name _____ Class _____ Date _____

3. **Draw Conclusions** Based on the graph, identify the days when each moon phase could have occurred: new moon, first quarter moon, full moon, last quarter moon. How do you know this?

   A new moon or a full moon could have occurred on Day 2 or 20. A new moon or a full moon could have occurred on Day 20. These days correspond with the highest tidal range. A first or third quarter moon could have occurred on Day 11 or 26. These days correspond with the lowest tidal range.

4. **Applying Concepts** On January 5th (Day 5 on the table) at 9:00 A.M., Jarred anchored his boat in about 4 feet of water at the beach. When he returned to his boat at 3:30 P.M., the boat was completely in the sand. What had happened? How long did Jarred have to wait to leave the area in his boat?

   The tide went out while Jarred was gone. At 3:25 P.M., low tide occurred. He would have to wait until the tide came in again so there would be enough water for him to move his boat. The longest he would have had to wait is about six hours, until high tide occurred at 9:32 P.M.

DATA TABLE Tidal Data for Long Beach, New York, January 2003

Day	Time*	Height*	Time	Height	Time	Height	Time	Height
1	05:45 A.M.	5.5	12:16 P.M.	−0.7	06:12 P.M.	4.4	—	—
2	12:18 A.M.	−0.5	06:35 A.M.	5.6	01:07 P.M.	−0.8	07:03 P.M.	4.4
3	01:10 A.M.	−0.5	07:23 A.M.	5.5	01:56 P.M.	−0.8	07:53 P.M.	4.4
4	01:59 A.M.	−0.4	08:11 A.M.	5.4	02:42 P.M.	−0.7	08:42 P.M.	4.3
5	02:45 A.M.	−0.2	08:59 A.M.	5.1	03:25 P.M.	−0.5	09:32 P.M.	4.2
6	03:30 A.M.	0.0	09:47 A.M.	4.8	04:07 P.M.	−0.3	10:23 P.M.	4.0
7	04:14 A.M.	0.3	10:35 A.M.	4.6	04:49 P.M.	−0.1	11:12 P.M.	3.9
8	05:01 A.M.	0.6	11:22 A.M.	4.3	05:32 P.M.	0.2	11:59 P.M.	3.9
9	05:54 A.M.	0.8	12:09 P.M.	4.0	06:18 P.M.	0.4	—	—
10	12:45 A.M.	3.9	06:56 A.M.	0.9	12:57 P.M.	3.7	07:10 P.M.	0.5
11	01:31 A.M.	3.9	07:59 A.M.	0.9	01:47 P.M.	3.5	08:02 P.M.	0.5
12	02:19 A.M.	4.0	08:57 A.M.	0.8	02:41 P.M.	3.4	08:53 P.M.	0.5
13	03:10 A.M.	4.1	09:50 A.M.	0.6	03:39 P.M.	3.5	09:41 P.M.	0.4
14	04:02 A.M.	4.3	10:38 A.M.	0.3	04:34 P.M.	3.6	10:28 P.M.	0.2
15	04:51 A.M.	4.6	11:26 A.M.	0.1	05:23 P.M.	3.7	11:15 P.M.	0.1
16	05:36 A.M.	4.8	12:12 P.M.	−0.1	06:08 P.M.	3.9	—	—
17	12:02 A.M.	−0.1	06:17 A.M.	5.0	12:57 P.M.	−0.3	06:51 P.M.	4.1
18	12:49 A.M.	−0.2	06:58 A.M.	5.1	01:40 P.M.	−0.5	07:32 P.M.	4.2
19	01:35 A.M.	−0.4	07:38 A.M.	5.2	02:22 P.M.	−0.6	08:15 P.M.	4.3
20	02:20 A.M.	−0.4	08:21 A.M.	5.2	03:30 P.M.	−0.7	09:01 P.M.	4.4
21	03:05 A.M.	−0.4	09:07 A.M.	5.1	03:44 P.M.	−0.7	09:51 P.M.	4.5
22	03:52 A.M.	−0.3	09:58 A.M.	4.9	04:27 P.M.	−0.6	10:44 P.M.	4.6
23	04:43 A.M.	−0.1	10:52 A.M.	4.7	05:13 P.M.	−0.4	11:37 P.M.	4.7
24	05:43 A.M.	0.1	11:48 A.M.	4.4	06:08 P.M.	−0.2	—	—
25	12:32 A.M.	4.7	06:53 A.M.	0.2	12:47 P.M.	4.2	07:11 P.M.	−0.1
26	01:30 A.M.	4.8	08:06 A.M.	0.2	01:50 P.M.	3.9	08:17 P.M.	0.0
27	02:31 A.M.	4.8	09:12 A.M.	0.1	02:57 P.M.	3.8	09:19 P.M.	0.0
28	03:35 A.M.	4.8	10:13 A.M.	−0.1	04:05 P.M.	3.9	10:17 P.M.	−0.1
29	04:37 A.M.	5.0	11:09 A.M.	−0.3	05:07 P.M.	4.0	11:13 P.M.	−0.2
30	05:33 A.M.	5.1	12:01 P.M.	−0.5	06:01 P.M.	4.2	—	—
31	12:06 A.M.	−0.3	06:22 A.M.	5.2	12:51 P.M.	−0.6	06:50 P.M.	4.3

*All times are listed in Local Standard Time (LST). All heights are in feet.
Source: Center for Operational Oceanographic Products and Services, National Oceanographic and Atmospheric Association, National Ocean Service.

Name _____ Class _____ Date _____

Chapter 17   The Atmosphere: Structure and Temperature   Exploration Lab
See the *Earth Science Teacher's Edition* for more information.

Heating Land and Water

The heating of Earth's surface controls the temperature of the air above it. To understand variations in air temperature, we consider the characteristics on the surface. Different land surfaces absorb varying amounts of incoming solar radiation. The largest contrast, however, is between land and water. The air temperature above water can influence the air temperature over land.

In this lab you will model the difference in the heating of land and water when they are subjected to a source of radiation. You first will assemble simple tools. Then you will observe and record temperature data. Finally, you will explain the results of the experiment and how they relate to the moderating influence of water on air temperatures near Earth's surface.

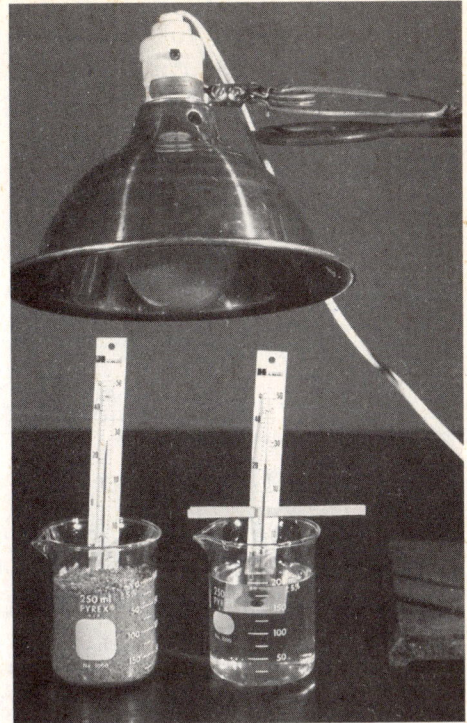

Figure 1

Problem How do the heating of land and water compare?

Materials
- 2 250-mL beakers
- dry sand
- tap water
- ring stand
- light source
- ruler
- 2 flat wooden sticks
- 2 thermometers
- United States map
- 3 different-colored pencils

Skills Modeling, Observing, Measuring, Analyzing Data

Procedure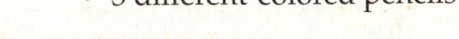

Part A: Preparing for the Experiment

1. Use the Data Table to record your measurements.
2. Pour 200 mL of dry sand into one of the beakers. Pour 200 mL of water into the other beaker.
3. Hang a light source from a ring stand so that it is about 5 inches above the beaker of sand and the beaker of water. The light should be situated so that it is at the same height above both beakers.
4. Using the wooden sticks, suspend a thermometer in each beaker, as shown in Figure 1. The thermometer bulbs should be just barely below the surfaces of the sand and the water.
5. Record the starting temperatures for both the dry sand and the water in the data table.

Part B: Heating the Beakers

CAUTION: *Do not touch the light source or the beakers without using thermal mitts.*

Name _____ Class _____ Date _____

6. Turn on the light. Observe and record the temperatures in the Data Table at one-minute intervals for 10 minutes.

7. Turn off the light for several minutes. Dampen the sand with water and record the starting temperature for damp sand. Repeat Step 6 for the damp sand.

DATA TABLE Land and Water Heating

	Starting Temperature	1 min	2 min	3 min	4 min	5 min	6 min	7 min	8 min	9 min	10 min
Water											
Dry sand											
Damp sand											

Analyze and Conclude

1. **Using Tables and Graphs** Use the data you collected to plot the temperatures for the water, dry sand, and damp sand. Use a different-colored line to connect the points for each material.

   Student graphs should have three different-colored lines on them: one representing water, another representing dry sand, and the last line representing damp sand. The graphs should indicate that dry sand heats up faster and gets hotter than water and damp sand.

2. **Comparing and Contrasting** How does the changing temperature differ for dry sand and water when they are exposed to equal amounts of radiation?

   Dry sand heats up faster and gets hotter than water does.

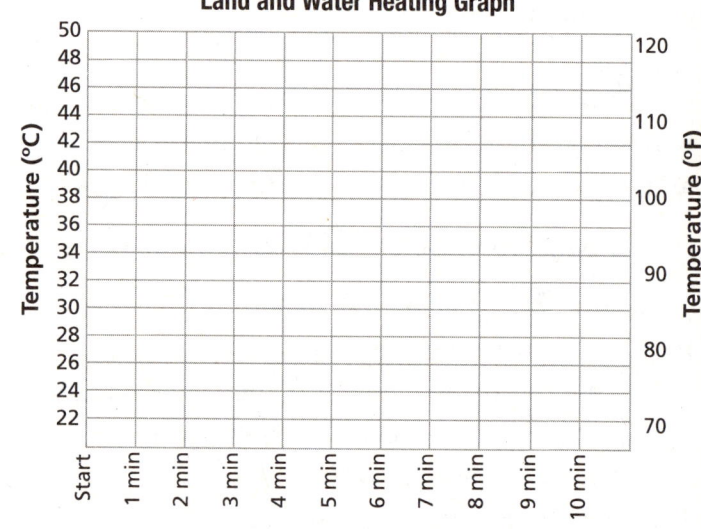
Land and Water Heating Graph

3. **Comparing and Contrasting** How does the changing temperature differ for dry sand and damp sand when they are exposed to equal amounts of radiation?

   Dry sand heats up and gets hotter faster than damp sand does.

4. **Applying** Locate Eureka, California, and Lafayette, Indiana, on a United States map. Infer which city would show the greater annual temperature range. Explain your answer.

   Lafayette, Indiana, would show the greater annual temperature range because of its continental location. There is no large body of water near it to cool off the land that absorbs more heat in the summer.

Name _____ Class _____ Date _____

Chapter 18   Moisture, Clouds, and Precipitation   Exploration Lab

Measuring Humidity

See the Earth Science Teacher's Edition for more information.

Relative humidity is a measurement used to describe water vapor in the air. In general, it expresses how close the air is to saturation. In this lab, you will use a psychrometer and a data table to determine the relative humidity of air.

Problem How can relative humidity be determined?

Materials
- calculator
- water at room temperature
- psychrometer

Alternative materials for psychrometer:
- 2 thermometers
- cotton gauze
- paper fan
- string

Skills Observing, Measuring, Analyzing Data, Calculating

Procedure

Part A: Calculating Relative Humidity from Water Vapor Content

1. Use Data Table 1 to record your measurements.
2. Relative humidity is the ratio of the air's water vapor content to its water vapor capacity at a given temperature. Relative humidity is expressed as a percent.

$$\text{Relative humidity (\%)} = \frac{\text{Water vapor content}}{\text{Water vapor capacity}} \times 100\%$$

3. At 25°C, the water vapor capacity is 20 g/kg. Use this information to complete Data Table 1.

DATA TABLE 1 Relative Humidity Determination Based on Water Vapor Content

Air temperature (°C)	Water Vapor Content (g/kg)	Water Vapor Capacity (g/kg)	Relative Humidity (%)
25	5	20	25
25	12	20	60
25	18	20	90

Name _____ Class _____ Date _____

Part B: Determining Relative Humidity Using a Psychrometer

4. A psychrometer consists of two thermometers—a wet-bulb thermometer and a dry-bulb thermometer. The wet-bulb thermometer has a cloth wick that is wet with water and spun for about 1 minute. Relative humidity is determined by calculating the difference in the temperature reading between the dry-bulb temperature and the wet-bulb temperature and using Data Table 2. For example, suppose a dry-bulb temperature is measured as 20°C, and a wet-bulb temperature is 14°C. Read the relative humidity from Data Table 2.

51 percent

5. If a psychrometer is not available, construct a wet-bulb thermometer by tying a piece of cotton gauze around the end of a thermometer. Wet it with room-temperature water, and fan it until the temperature stops changing.

6. Make wet-bulb and dry-bulb temperature measurements for the air in your classroom and the air outside. Use Data Table 3 to record your measurements. Use your measurements and Data Table 2 to determine the relative humidity inside and outside.

DATA TABLE 2 Relative Humidity (percent)

Dry-Bulb Temperature (°C)	Depression of Wet-Bulb Temperature (Dry-Bulb Temperature − Wet-Bulb Temperature = Depression of the Wet Bulb)																					
	1	2	3	4	5	6	7	8	9	10	11	12	13	14	15	16	17	18	19	20	21	22
−20	28																					
−18	40																					
−16	48	0																				
−14	55	11																				
−12	61	23																				
−10	66	33	0																			
−8	71	41	13																			
−6	73	48	20	0																		
−4	77	54	43	11																		
−2	79	58	37	20	1																	
0	81	63	45	28	11																	
2	83	67	51	36	20	6																
4	85	70	56	42	27	14																
6	86	72	59	46	35	22	10	0														
8	87	74	62	51	39	28	17	6														
10	88	76	65	54	43	33	24	13	4													
12	88	78	67	57	48	38	28	19	10	2												
14	89	79	69	60	50	41	33	25	16	8	1											
16	90	80	71	62	54	45	37	29	21	14	7	1										
18	91	81	72	64	56	48	40	33	26	19	12	6	0									
20	91	82	74	66	58	51	44	36	30	23	17	11	5	0								
22	92	83	75	68	60	53	46	40	33	27	21	15	10	4	0							
24	92	84	76	69	62	55	49	42	36	30	25	20	14	9	4	0						
26	92	85	77	70	64	57	51	45	39	34	28	23	18	13	9	5						
28	93	86	78	71	65	59	53	47	42	36	31	26	21	17	12	8	2					
30	93	86	79	72	66	61	55	49	44	39	34	29	25	20	16	12	8	4				
32	93	86	80	73	68	62	56	51	46	41	36	32	27	22	19	14	11	8	4			
34	93	86	81	74	69	63	58	52	48	43	38	34	30	26	22	18	14	11	8	5		
36	94	87	81	75	69	64	59	54	50	44	40	36	32	28	24	21	17	13	10	7	4	
38	94	87	82	76	70	66	60	55	51	46	42	38	34	30	26	23	20	16	13	10	7	5
40	94	89	82	76	71	67	61	57	52	48	44	40	36	33	29	25	22	19	16	13	10	7

Name _____ Class _____ Date _____

DATA TABLE 3 Relative Humidity Determinations Using Dry- and Wet-Bulb Thermometers

	Inside	Outside
Dry-bulb temperature (°C)		
Wet-bulb temperature (°C)		
Differences between dry-bulb and wet-bulb temperatures (°C)		
Relative humidity (%)		

Analyze and Conclude

1. **Comparing and Contrasting** How do the relative humidity measurements for inside and outside compare? Why are your determinations similar or different?

   Outdoor relative humidity determinations can be compared to meteorological reports for the day. Indoor relative humidity will vary widely depending on whether classrooms are air-conditioned, what season it is, or other variables.

2. **Applying Concepts** Explain the principle behind using a psychrometer to determine relative humidity.

   Relative humidity determinations with the psychrometer depend on the evaporation rate of the wick on the wet bulb. The dryer the air, the larger the wet-bulb temperature will be depressed, and the lower the relative humidity.

3. **Applying Concepts** Suppose you hear on the radio that the relative humidity is 90 percent on a winter day. Can you conclude that this air contains more moisture than air on a summer day with a 40 percent relative humidity? Explain why or why not.

   No; relative humidity depends on the capacity of air to hold water, which varies with temperature. The actual water content of winter air with 90 percent relative humidity can be less than summer air with 40 percent relative humidity.

4. **Applying Concepts** Why is a cool basement often damp in the summer?

   The relative humidity increases as air is cooled in the basement.

Earth Science Lab Manual

Name _____ Class _____ Date _____

Chapter 19   Air Pressure and Wind   **Exploration Lab**

Observing Wind Patterns

See the *Earth Science Teacher's Edition* for more information.

Atmospheric pressure and wind are two elements of weather that are closely interrelated. Most people don't usually pay close attention to the pressure given in a weather report. However, pressure differences in the atmosphere drive the winds that often bring changes in temperature and moisture.

Problem How can surface barometric pressure maps be interpreted?

Materials
- Resource 21 in the DataBank

Skills Observing, Analyzing Data, Calculating

Procedure
1. Look at Figure 2 on the next page. This map shows global wind patterns and average global barometric pressure for the month of January.
2. Examine the individual pressure cells—the isobars around the letters H and L—in Figure 2. Then complete the diagrams in your copy of Figure 1. Label the isobars with appropriate pressures, and use arrows to indicate the surface air movement in each pressure cell.
3. Indicate the movements of air in high and low pressure cells by completing the Data Table below.

Northern Hemisphere

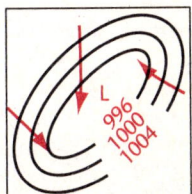

Southern Hemisphere

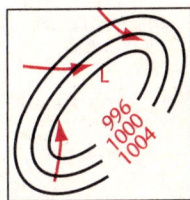

Figure 1

DATA TABLE Air Movements in Pressure Cells

Air Movement	N. Hem. High	N. Hem. Low	S. Hem. High	S. Hem. Low
Into/out of	out of	into	out of	into
Rises/sinks	sinks	rises	sinks	rises
Rotates CW/CCW*	CW	CCW	CCW	CW

*CW = clockwise; CCW = counterclockwise

Earth Science Lab Manual ▪ 215

Name _____ Class _____ Date _____

Figure 2

Analyze and Conclude

1. **Comparing and Contrasting** Summarize the differences and similarities in surface air movement between a Northern Hemisphere cyclone and a Southern Hemisphere cyclone.

   In a Northern Hemisphere cyclone, air rotates counterclockwise, whereas air rotates clockwise in an anticyclone. These rotation directions result from the Coriolis effect, and they are reversed for cyclones and anticyclones in the Southern Hemisphere. In either hemisphere, the net flow of air is inward for a cyclone and outward for an anticyclone. This net effect inward and outward is due to friction. Also, air sinks in an anticyclone and rises in a cyclone in both hemispheres.

2. **Interpreting Illustrations** Use Resource 21 in the DataBank as a reference to locate and write the name of each global wind belt at the appropriate location on the map in Figure 2. Also indicate the region of the polar front. Student locations of wind belts, pressure zones, and fronts should be like those in Resource 21 in the DataBank.

3. **Applying** Label the areas on Figure 2 where you would expect high wind speeds to occur. Areas of high wind speed correlate to closely spaced isobars. Students should indicate these areas on their maps.

4. **Applying** Label areas on Figure 2 where circulation is most like the idealized global wind model for a rotating Earth. Explain why this region on Earth is so much like the model.

   Students should locate the subpolar low in the Southern Hemisphere, at approximately 60 degrees south latitude. There are no landmasses in the region of this subpolar low to create seasonal temperature differences. Therefore, the air circulation pattern is not disrupted.

Earth Science Lab Manual • 216

Name _____ Class _____ Date _____

Chapter 20 Weather Patterns and Severe Storms

Application Lab

See the *Earth Science Teacher's Edition* for more information.

Middle-Latitude Cyclones

You've learned that much of the day-to-day weather in the United States is caused by middle-latitude cyclones. In this lab, you will identify some of the atmospheric conditions associated with a middle-latitude cyclone. Then you will use what you know about Earth's atmosphere and weather to predict how the movement of the low-pressure system affects weather in the area.

Problem How do middle-latitude cyclones affect weather patterns?

Materials
- colored pencils
- Resource 22 in the DataBank

Skills Observing, Comparing and Contrasting, Predicting

Procedure

1. Use the colored pencils to color the land and water areas on the map in Figure 1. Also color the symbols used to designate the fronts.

2. Identify and label the cold front, warm front, and occluded front on the map in Figure 1.

3. Draw arrows that show the direction of surface winds at points A, C, E, F, and G.

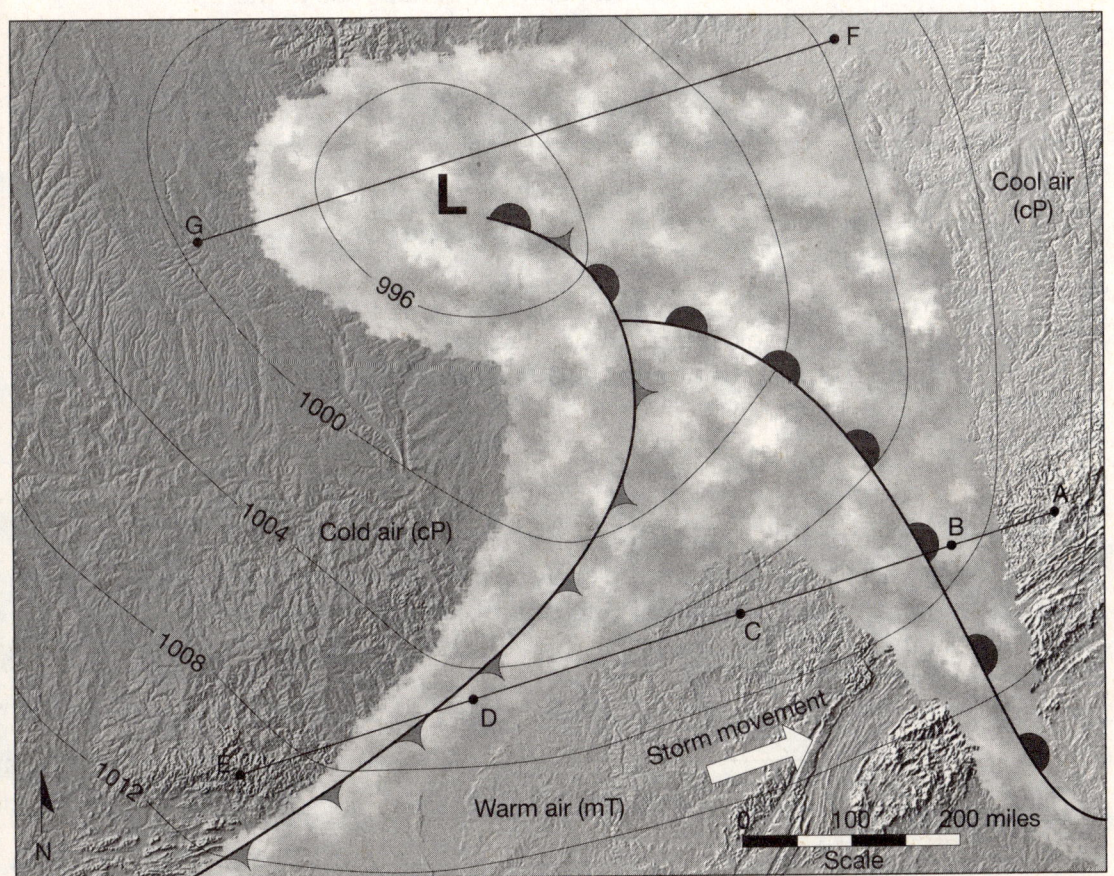

Figure 1

Earth Science Lab Manual ▪ 217

Name _____ Class _____ Date _____

Analyze and Conclude

1. **Describing** In which direction are the surface winds moving?

   The winds at point A are blowing toward north-northwest. At point C, they are blowing almost

   directly north. At point E, the winds are moving almost directly east. At point F, the winds are blowing

   from the southeast toward the northwest. At point G, the winds are blowing from the northwest to

   the southeast.

2. **Identifying** At which stage of formation is the cyclone? Explain your answer. Refer to the map on Resource 22 in the DataBank if necessary.

   The middle-latitude cyclone shown is at the mature stage as indicated by the occluded front.

3. **Explaining** Is the air in the center of the cyclone rising or falling? What effect does this have on the potential for condensation and precipitation?

   The air at the center of the cyclone is rising. The potential for condensation and precipitation will

   be good because as the air rises, cooling will occur and the dew-point temperature may be reached.

4. **Inferring** Find the center of the low, which is marked with the letter *L*. What type of front has formed here? What happens to the maritime tropical air in this type of front?

   An occluded front has formed at the center of the low pressure system. The warm maritime tropical

   air is lifted above the cooler air.

5. **Predicting** Once the warm front passes, in which direction will the wind at point B blow?

   After the warm front passes, the wind at point B will be from the south.

6. **Synthesizing** Describe the changes in wind direction and moisture in the air that will likely occur at point D after the cold front passes.

   The wind will blow from the northwest, and the barometric pressure will rise at point D after the

   cold front passes.

7. **Synthesizing** Describe the wind directions, humidity, and precipitation expected for a city as the cyclone moves and the city's relative position changes from point A to point B, point C, point D, and finally from point D to point E.

   point A to point B: low stratus clouds, possible precipitation, wind from the southeast, barometric

   pressure falling; point C: wind from the south, warm and perhaps humid air; point D: vertical clouds,

   possible thunderstorms; point D to point E: barometric pressure rising, wind from the northwest, cool

   or cold temperatures, clearing sky

Chapter 21 Climate

Exploration Lab

Human Impact on Climate and Weather

See the *Earth Science Teacher's Edition* for more information.

Scientists are now closely monitoring how daily human activity is changing microclimates. There is concern that changing microclimates can have an effect on global climates. In this investigation, you will explore some of the ways that human activities are changing the atmosphere.

Problem How do we know that human activity is changing Earth's climates?

Materials
- paper
- pen or pencil

Skills Calculating, Measuring, Using Tables, Analyzing Data

Procedure

1. Data Table 1 lists many of the types, sources, and amounts of primary pollutants. Use this table to answer Questions 1, 2, 3, and 4 under Analyze and Conclude.

DATA TABLE 1 Estimated Nationwide Emissions (millions of metric tons/year)

Source	Carbon Monoxide	Particulates	Sulfur Oxides	Volatile Organics	Nitrogen Oxides	Total
Transportation	43.5	1.6	1.0	5.1	7.3	58.5
Stationary source fuel combustion	4.7	1.9	16.6	0.7	10.6	34.5
Industrial processes	4.7	2.6	3.2	7.9	0.6	19.0
Solid waste disposal	2.1	0.3	0.0	0.7	0.1	3.2
Miscellaneous	7.2	1.2	0.0	2.8	0.2	11.4
Total	62.2	7.6	20.8	17.2	18.8	126.6

Source: U.S. Environmental Protection Agency

2. Look at Figure A. The pollutants listed are linked to a wide variety of negative health effects such as eye irritation, heart damage, and lung damage. The pollutants shown are also linked to reduced visibility, reduced crop yields, and damage to ecosystems. Study the figure and answer Questions 5, 6, and 7.

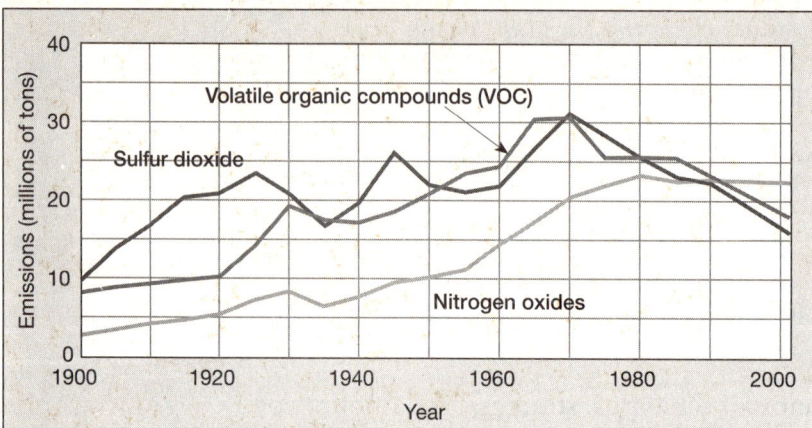

Figure A

3. Look at Figure B. Scientists have noted the increasing levels of carbon dioxide in the atmosphere. Research continues to determine whether these increasing levels are affecting global climates. Use Figure B to answer Question 8.

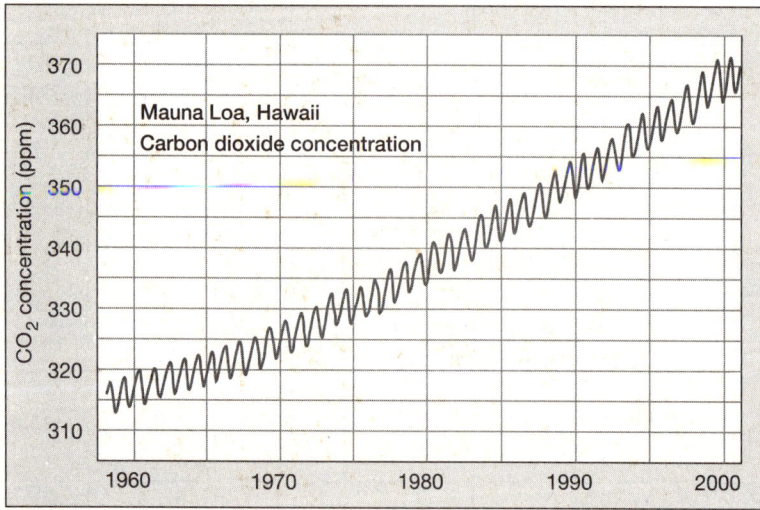

Figure B

Earth Science Lab Manual ▪ 220

4. Look at Data Table 2. This table presents data on the effects of large cities on their surrounding microclimates. Summer temperatures in cities can be higher than the surrounding countryside. Meteorologists call this effect "the urban heat island." Study the data in the table and answer Questions 9, 10, and 11.

DATA TABLE 2 Average Climatic Changes Produced by Cities

Element	Comparison with Rural Temperature
Particulate matter	10 times more
Temperature	
Annual mean	0.5–1.5°C higher
Winter	1–2°C higher
Solar radiation	15–30% less
Ultraviolet, winter	30% less
Ultraviolet, summer	5% less
Precipitation	5–15% more
Thunderstorm frequency	16% more
Winter	5% more
Summer	29% more
Relative humidity	6% lower
Winter	2% lower
Summer	8% lower
Cloudiness (frequency)	5–10% more
Fog (frequency)	60% more
Winter	100% more
Summer	30% more
Wind speed	25% lower
Calms	5–20% more

Source: After Landsberg, Changnon, and others

Analyze and Conclude

1. **Interpreting Data** What is the leading source (by weight) of primary pollutants? How many metric tons of this pollutant are added to the atmosphere each year?

   transportation; 58.5 million metric tons

2. **Interpreting Data** Which of the following is the most abundant primary pollutant?
   a. carbon monoxide
   b. sulfur oxides

   a. carbon monoxide

3. **Calculating** Your answer for item 2 is what percentage of all primary pollutants?
   a. 25%   b. 50%   c. 75%

   b. 50%

4. **Calculating** What is the approximate total weight (in million metric tons) of all primary pollutants added to the atmosphere?

   126.6 million metric tons

5. **Interpreting Data** Describe the trend you see in the data for atmospheric pollutants prior to 1970.

   Emissions of atmospheric pollutants were generally on the rise prior to 1970.

6. **Interpreting Data** Describe the trend you see in the data for atmospheric pollutants since 1970.

   Atmospheric pollutant emissions have generally decreased since 1970.

Name _____ Class _____ Date _____

7. **Inferring** Suggest a reason for the changing trends in Questions 5 and 6.

   Industrialization and increased transportation using fossil fuels increased CO_2 concentrations until 1970, when government regulations took effect and began lowering CO_2 concentrations.

8. **Calculating** What has been the approximate percentage increase in atmospheric carbon dioxide near Mauna Loa since 1958?

   net change: 370 ppm − 315 ppm = 55 ppm; 55 ppm ÷ 370 ppm × 100 = approximately 15 percent increase in atmospheric carbon dioxide

9. **Interpreting Data** Compared to rural areas, which factors are increased by urbanization? Which factors are decreased?

   Increased: particulate matter, temperature, precipitation, thunderstorm frequency, cloudiness, fog, calms; decreased: solar radiation, relative humidity, wind speed

10. **Interpreting Data** Of all of the factors shown, which shows the greatest increase due to urbanization?

    fog in winter

11. **Predicting** Suggest a possible reason for each of the following effects on the weather that is influenced by a city.

    a. increased frequency of thunderstorms
    b. lower wind speed
    c. increased precipitation

    a. Urban heating reduces the stability of the atmosphere.

    b. Buildings impede the flow of air.

    c. Urban heating reduces the stability of the atmosphere; there are more condensation nuclei from industrial pollution; buildings impede the movement of weather systems causing rain-producing weather to linger.

Name _____ Class _____ Date _____

Chapter 22 Origin of Modern Astronomy

Exploration Lab

Modeling Synodic and Sidereal Months

See the *Earth Science Teacher's Edition* for more information.

The time interval required for the moon to complete a full cycle of phases is 29.5 days, or one synodic month. The true period of the moon's revolution around Earth, however, takes only 27.3 days and is known as the sidereal month. In this lab, you will model the differences between synodic and sidereal months.

Problem How do synodic and sidereal months differ?

Materials
- lamp
- basketball
- softball

Skills Observing, Using Models, Analyzing Data, Drawing Conclusions

Procedure

1. On the diagram of Month 1, indicate the dark half of the moon on each of the eight lunar positions by shading the appropriate area with a pencil.

2. On the diagram of Month 1, label the position of the new moon. Do the same for the other lunar phases.

3. Repeat Steps 1 and 2 for the diagram of Month 2.

4. Place the lamp on a desk or table. The lamp represents the sun. Hold the softball, which represents the moon. Have a partner hold the basketball, which represents Earth.

5. Stand so that the "moon" is in the position of the new-moon phase in Month 1, relative to "Earth" and the "sun." Revolve the moon around Earth while at the same time moving both Earth and the moon to Month 2. Stop at the same numbered position at which you began. Use the diagrams to guide your movements.

Analyze and Conclude

1. **Using Models** After one complete revolution beginning at the new-moon phase in Month 1, in what position is the moon located in Month 2?

   position 6

2. **Interpreting Data** Based on your answer to the previous question, does this position occur before or after the moon has completed one full cycle of phases?

   before

Name _____ Class _____ Date _____

3. Identifying In Month 2, what position represents the new-moon phase? When the moon reaches this position, will it have completed a synodic or sidereal month?

position 7; a synodic month

4. Summarizing In your own words, explain the difference between a sidereal and synodic month.

A sidereal month is one complete revolution of the moon around Earth. During a synodic month, the moon completes one cycle of phases. The synodic month is about two days longer than the sidereal month because of the motion of the Earth-moon system around the sun.

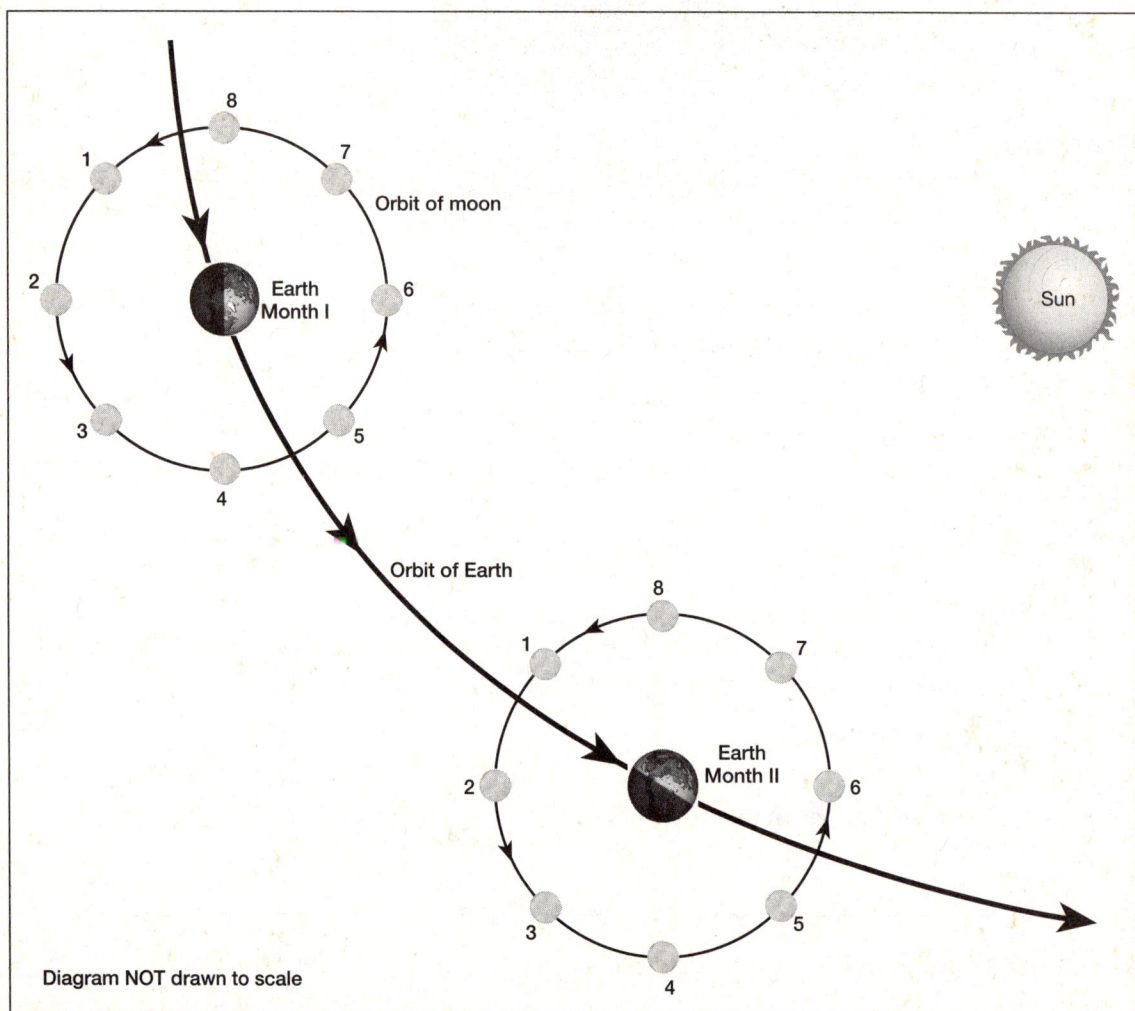

Figure 1

Name _____ Class _____ Date _____

Chapter 23 Touring Our Solar System

Exploration Lab

Modeling the Solar System

See the Earth Science Teacher's Edition for more information.

An examination of any scale model of the solar system reveals that the distances from the sun and the spacing between the planets appear to follow a regular pattern. The best way to examine this pattern is to build an actual scale model of the solar system.

Problem How can you model distances among the planets and their distances from the sun?

Materials

- meter stick
- calculator
- colored pencils
- 6-meter length of adding machine paper

Skills Calculating, Using Models

Procedure

Note: Use Figure 1 on the next page to help you model the solar system.

1. Place the 6-meter length of adding machine paper on the floor.
2. Draw an X about 10 centimeters from one end of the adding machine paper. Label this mark "sun."
3. The Data Table shows the mean distances of the planets from the sun, as well as their diameters. Use the table and the following scale to calculate the proper scale distance of each planet from the sun:

   1 millimeter = 1 million kilometers

   1 centimeter = 10 million kilometers

   1 meter = 1000 million kilometers

4. After calculating the scale distances, draw on a separate sheet of paper a small circle for each planet at its proper scale distance from the sun. Use a different-colored pencil for the inner and outer planets. Write the name of each planet next to its position.

DATA TABLE

Planet	Distance from Sun		Diameter (km)	Scale Distance from Sun
	AU	Millions of km		
Mercury	0.39	58	4878	
Venus	0.72	108	12,104	
Earth	1.00	150	12,756	
Mars	1.52	228	6794	
Jupiter	5.20	778	143,884	
Saturn	9.54	1427	120,536	
Uranus	19.18	2870	51,118	
Neptune	30.06	4497	50,530	
Pluto	39.44	5900	2300	

Analyze and Conclude

1. **Using Models** How far from the sun is Earth located on your model? How far from the sun are the rest of the planets located?

   Earth: 15 cm; Mercury: 5.8 cm; Venus: 10.8 cm; Mars: 22.8 cm; Jupiter: 77.8 cm;

   Saturn: 1.4 m; Uranus: 2.9 m; Neptune: 4.5 m; Pluto: 5.9 m

2. **Observing** What pattern of spacing do you observe? Summarize the pattern for both the inner and outer planets.

   The inner planets are closely spaced together. The largest space exists between

   Earth and Mars. The outer planets are widely spaced apart. The smallest space

   is between Jupiter and Saturn.

3. **Interpreting Data** Which planet or planets vary most from the general pattern of spacing?

   The space between Neptune and Pluto varies the most from the general pattern.

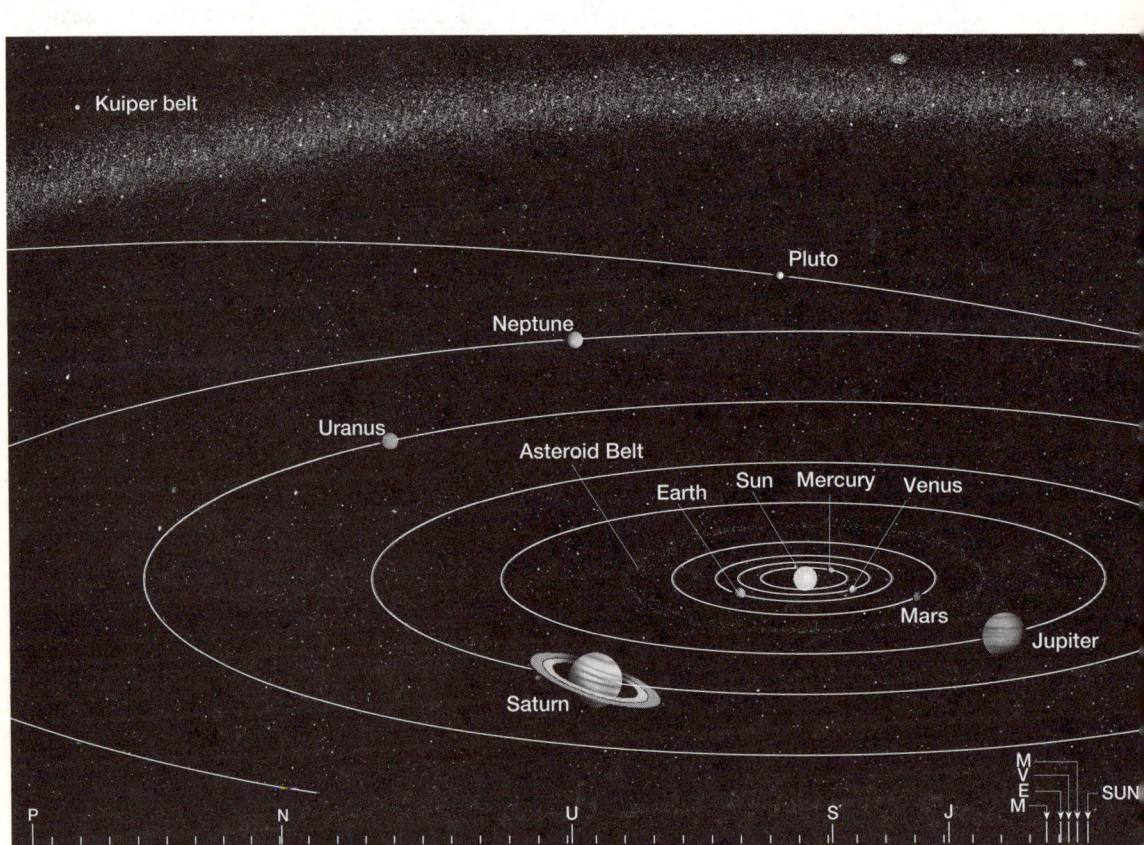

Figure 1

Name _____ Class _____ Date _____

Chapter 24 Studying the Sun

Exploration Lab

Tracking Sunspots

See the *Earth Science Teacher's Edition* for more information.

Sunspots begin as small areas about 1600 kilometers in diameter. Most last for only a few hours. However, some grow into dark regions many times larger than Earth and last for a month or more. In this lab you will count the number of sunspots over the course of several days.

Problem How can you use a telescope to safely view and count the number of sunspots on the sun's surface?

Materials
- telescope
- large cardboard box
- metric ruler
- small cardboard box
- piece of white paper
- tape

Skills Observing, Interpreting Data, Making and Using Graphs

Procedure 🛠️ ⚠️ *Caution students to follow the setup and procedure exactly to avoid damaging one's eyesight or the equipment.*

⚠️ 1. Position a telescope on a tripod outside in a sunny spot away from trees and other obstacles. The eyepiece should face away from the sun. **CAUTION:** *Never look at the sun directly. Do not view the sun through the telescope. These actions could cause eye damage.*

2. Place the large cardboard box on the ground about 15 centimeters in front of the telescope's eyepiece.

3. Use the pencil to punch a hole in one side of the small cardboard box. Tape a sheet of white paper inside the opposite end of the box, as shown in the illustration.

4. Place the small box on top of the large box so that its front is open for viewing. The hole in the small box should face the eyepiece of the telescope. Adjust the telescope so that the eyepiece, the hole, and the white paper are aligned.

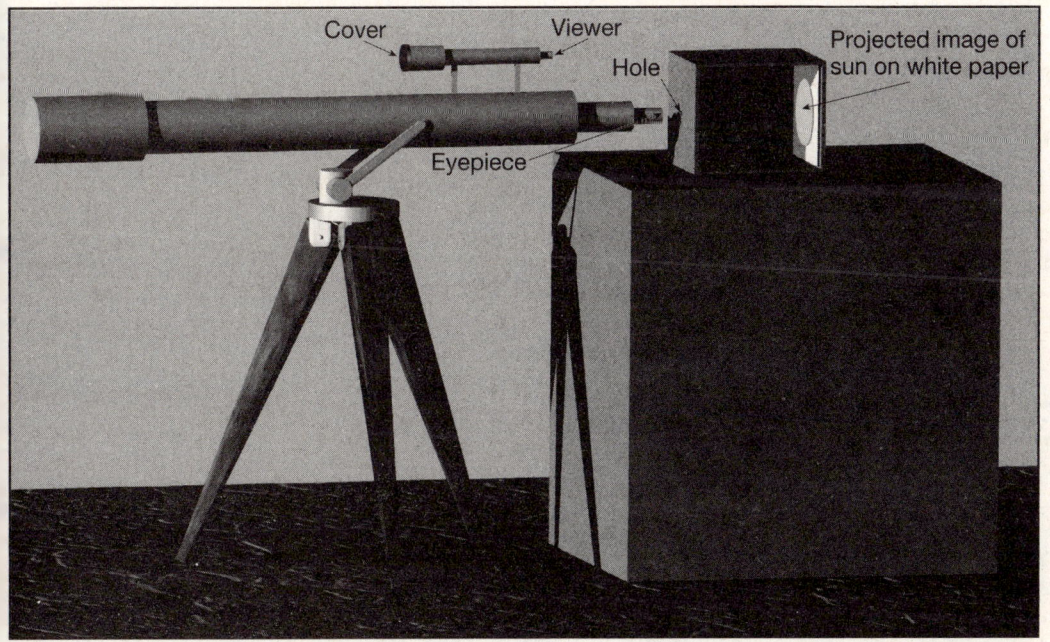

Earth Science Lab Manual ▪ **227**

Name _____ Class _____ Date _____

5. Adjust the small box until you see an image of the sun projected onto the paper. You may adjust the telescope to obtain a clearer image, but do not look through the viewer to accomplish this. You may also vary the distance between the box and the telescope to obtain better images.

6. Record the number of sunspots that you observe in the Data Table. Trace the outlines of sunspots on the paper in the box. Shade in the sunspots and use the ruler to measure their size.

7. As weather permits, make several more viewings of sunspots over the course of the next few days. During each viewing, repeat Steps 1–6. Be sure to note the movement of the sunspots in the Data Table.

DATA TABLE

Day	Number of Sunspots	Movement?
1		
2		
3		
4		
5		

Analyze and Conclude

1. **Making Graphs** How many sunspots did you observe? Make a line graph of the data from your data table.

   Sample answer: about 15 large sunspots;

   check students' graphs.

2. **Observing** How did the number of sunspots vary over the course of your observations?

   Sample answer: The number of sunspots

   remained fairly steady over the course

   of the observations.

3. **Interpreting Data** Why did the sunspots move?

   The sunspots moved because the sun rotates on its axis.

Earth Science Lab Manual ▪ 228

Name _____ Class _____ Date _____

Chapter 25 Beyond Our Solar System   Exploration Lab

Observing Stars

See the *Earth Science Teacher's Edition* for more information.

Throughout history, people have been recording the nightly movement of stars that results from Earth's rotation, as well as the seasonal changes in the constellations as Earth revolves around the sun. Early astronomers offered many explanations for the changes before the true nature of the motions was understood in the seventeenth century. In this lab, you'll observe and identify stars.

Problem How can you use star charts to identify constellations and track star movements?

Materials
- Resource 19 in the DataBank
- Resource 21 in the DataBank
- penlight
- notebook

Skills Observing, Summarizing, Interpreting Data

Procedure
1. On a clear, moonless night far from street lights, go outside and observe the stars.
2. In the Data Table, record the date and make a list of the different colors of stars that you see.
3. Select one star that is overhead or nearly so. Observe and record its movement over a period of one hour. Also note the direction of its movement (eastward, westward).

DATA TABLE

Date	Star Colors	Star Movement	Constellations	Motions of Stars Around North Star

Name _____ Class _____ Date _____

4. Select a star chart suitable for your location and season. Locate and record the names of several constellations that you see in the sky. Sketch and label them in the box below.

5. Locate the North Star (Polaris) in the night sky. Observe and record the motion of stars that surround the North Star.
6. Repeat your observations several weeks later at the same location.

Name _____ Class _____ Date _____

Analyze and Conclude

1. **Observing** How many different colors of stars did you observe? How do these colors relate to star temperature? *Hint:* The Hertzsprung-Russell diagram in the Appendix may help you.

    Observed colors were blue, blue-white, red, and yellow. Blue and blue-white stars are hot,

    red stars are cool, and yellow stars are of medium temperature.

2. **Interpreting Data** In which direction did the star that you observed appear to move? How is this movement related to the direction of Earth's rotation?

    The star appeared to move westward. The rotation of Earth from west to east (eastward) makes

    the position of the star appear to move east to west (westward) throughout the night.

3. **Summarizing** Write a brief summary of the motion of the stars that surround the North Star. Be sure to include any changes you observed during your second viewing.

    The stars surrounding the North Star appear to move in circles around the North Star, with the

    circles becoming larger the farther a star appears to be from the North Star.

Resource 1: Map Symbols

Control data and monuments
Vertical control
- Third order or better, with tablet — BM ×16.3
- Third order or better, recoverable mark — ×120.0
- Bench mark at found section corner — BM ×18.6
- Spot elevation — ×5.3

Contours
Topographic
- Intermediate
- Index
- Supplementary
- Depression
- Cut; fill

Bathymetric
- Intermediate
- Index
- Primary
- Index primary
- Supplementary

Boundaries
- National
- State or territorial
- County or equivalent
- Civil township or equivalent
- Incorporated city or equivalent
- Park, reservation, or monument

Surface features
- Levee
- Sand or mud area, dunes, or shifting sand (Sand)
- Intricate surface area (Strip mine)
- Gravel beach or glacial moraine (Gravel)
- Tailings pond (Tailings pond)

Mines and caves
- Quarry or open pit mine
- Gravel, sand, clay, or borrow pit
- Mine dump (Mine dump)
- Tailings (Tailings)

Vegetation
- Woods
- Scrub
- Orchard
- Vineyard
- Mangrove

Glaciers and permanent snowfields
- Contours and limits
- Form lines

Marine shoreline
Topographic maps
- Approximate mean high water
- Indefinite or unsurveyed

Topographic-bathymetric maps
- Mean high water
- Apparent (edge of vegetation)

Coastal features
- Foreshore flat
- Rock or coral reef
- Rock bare or awash
- Group of rocks bare or awash
- Exposed wreck
- Depth curve; sounding
- Breakwater, pier, jetty, or wharf
- Seawall

Rivers, lakes, and canals
- Intermittent stream
- Intermittent river
- Disappearing stream
- Perennial stream
- Perennial river
- Small falls; small rapids
- Large falls; large rapids
- Masonry dam
- Dam with lock
- Dam carrying road
- Perennial lake; Intermittent lake or pond
- Dry lake (Dry lake)
- Narrow wash
- Wide wash
- Canal, flume, or aquaduct with lock
- Well or spring; spring or seep

Submerged areas and bogs
- Marsh or swamp
- Submerged marsh or swamp
- Wooded marsh or swamp
- Submerged wooded marsh or swamp
- Rice field (Rice)
- Land subject to inundation (Max pool 431)

Buildings and related features
- Building
- School; church
- Built-up area
- Racetrack
- Airport
- Landing strip
- Well (other than water); windmill
- Tanks
- Covered reservoir
- Gaging station
- Landmark object (feature as labeled)
- Campground; picnic area
- Cemetery: small; large (Cem)

Roads and related features
Roads on Provisional edition maps are not classified as primary, secondary, or light duty. They are all symbolized as light duty roads.
- Primary highway
- Secondary highway
- Light duty road
- Unimproved road
- Trail
- Dual highway
- Dual highway with median strip

Railroads and related features
- Standard gauge single track; station
- Standard gauge multiple track
- Abandoned

Transmission lines and pipelines
- Power transmission line; pole; tower
- Telephone line — Telephone
- Aboveground oil or gas pipeline
- Underground oil or gas pipeline — Pipeline

Symbols used on topographic quadrangle maps produced by the U.S. Geological Survey. Variations will be found on older maps.

DataBank

Earth Science Lab Manual · DB1

Resource 2 — Identifying Crystal Systems

Crystal A

Crystal B

Crystal D

Crystal C

Crystal E

Crystal F

Crystal G

Crystal H

Crystal J

Crystal I

Earth Science Lab Manual

Resource 3: Earth's Tectonic Plates

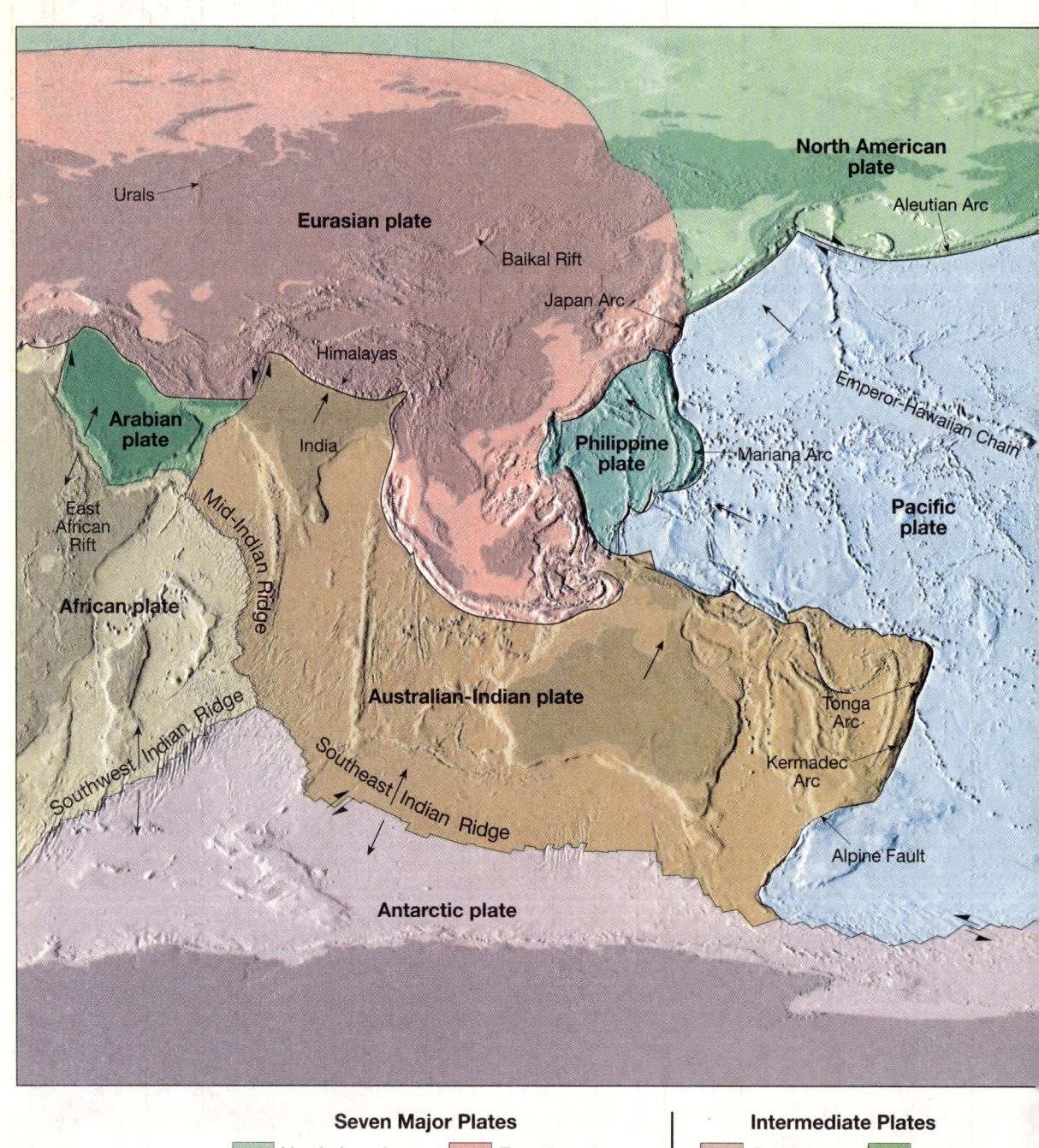

Resource 4: Global Bathymetry from Altimetry Data

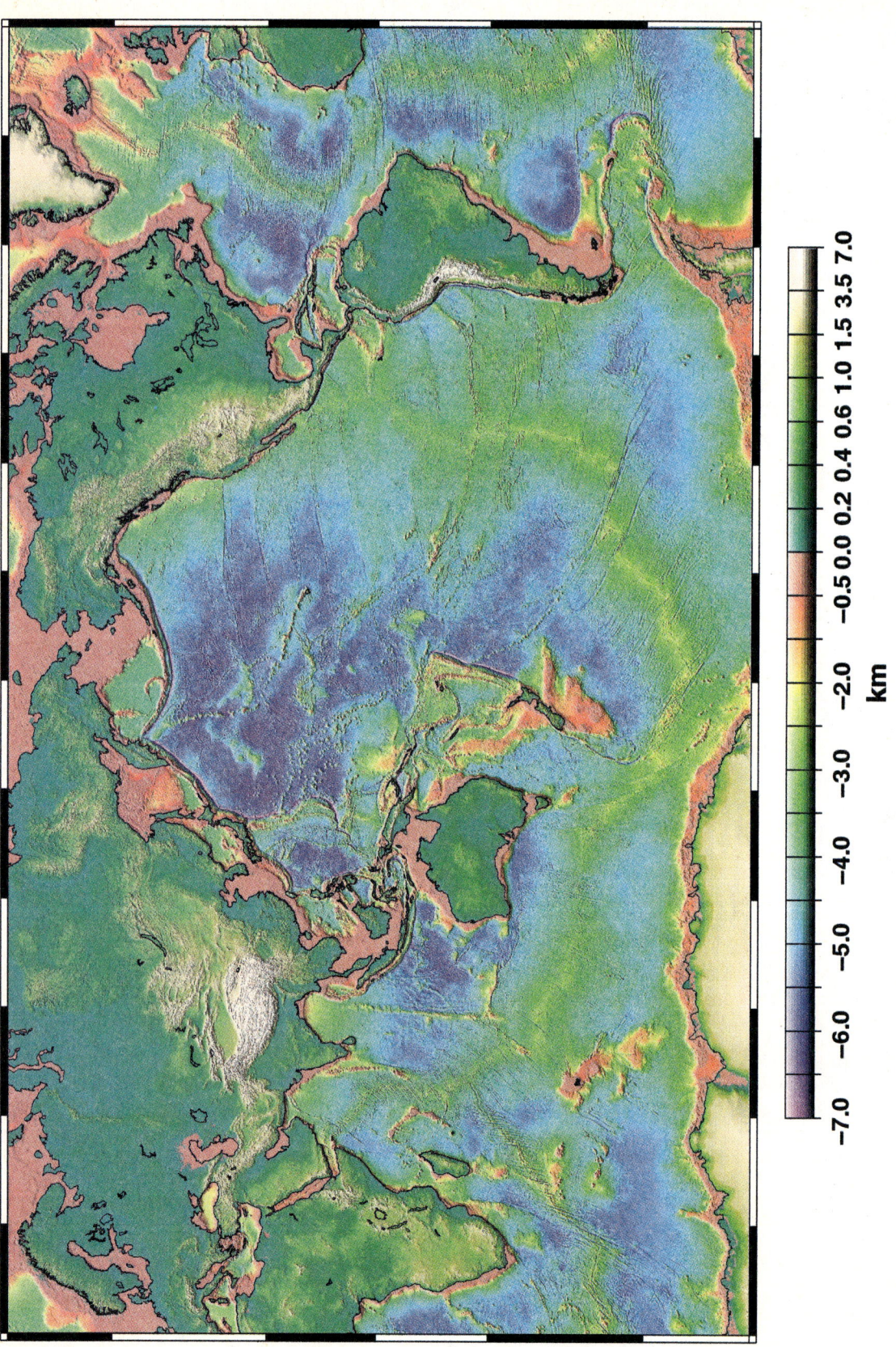

Earth Science Lab Manual ▪ **DB6**

Resource 5: Ridge Fracture Zone

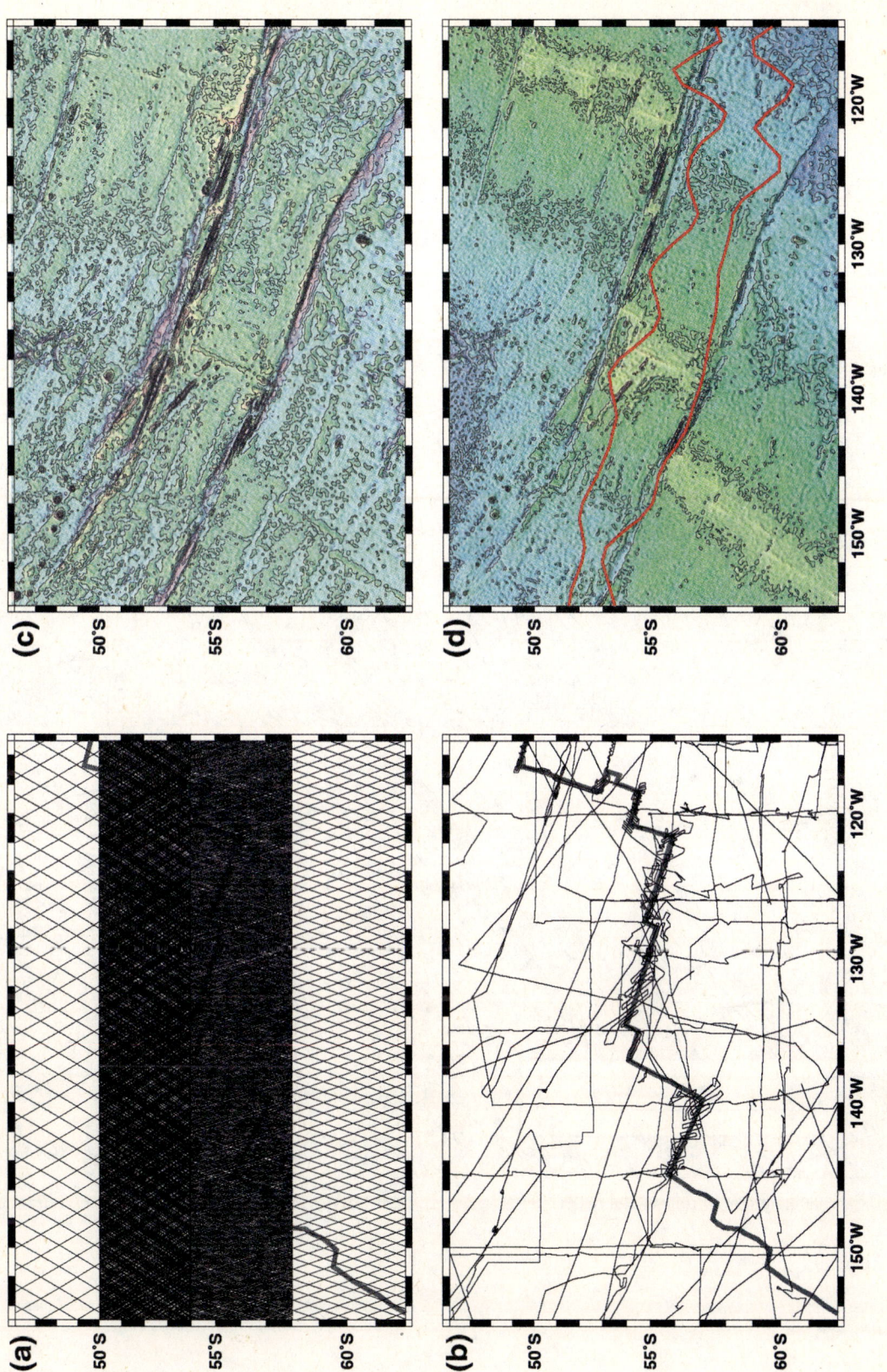

Earth Science Lab Manual ▪ DB7

Resource 6 — Geologic Map of Devil's Fence, Montana

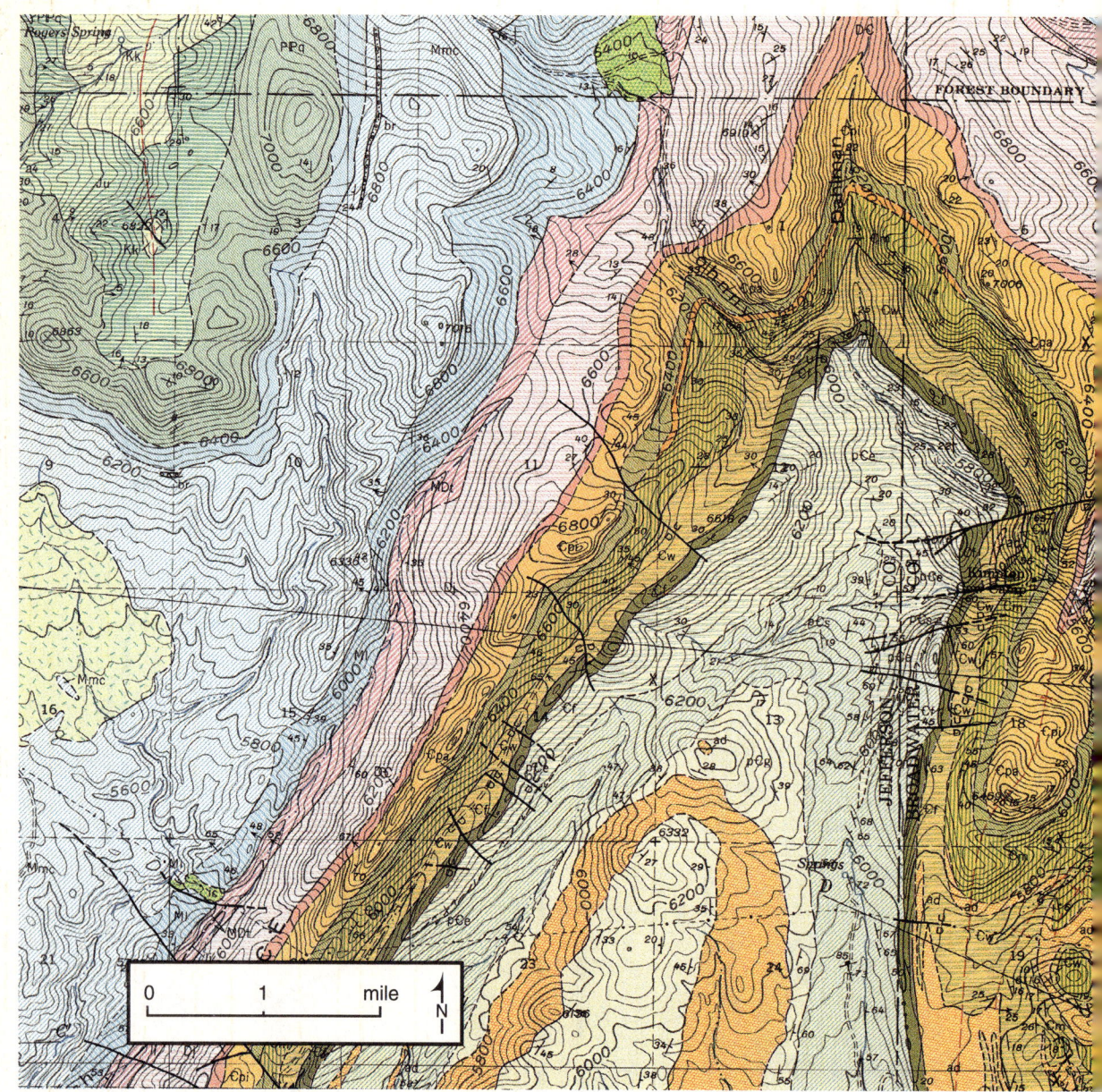

Source: U.S. Geologic Survey, Professional Paper 292

SCALE 1:31250
CONTOUR INTERVAL 40 FEET
DATUM IS MEAN SEA LEVEL

Resource 7 — Map Key for Geologic Map

EROSIONAL UNCONFORMITY

Phosphoria formation
Brown and gray chert and sandstone, in part phosphatic; may locally contain one or two thin beds of phosphate rock, Pp. In places mapped with the Quadrant formation, PPq.

Quadrant formation
Light-colored quartzitic sandstone and interbedded light-gray sugary-textured sandy dolomite

Amsden formation
Red to grayish-red mudstone, shale, and subordinate amounts of carbonate rock with interbeds of gray, brown, or yellow argillaceous sandstone in upper and lower parts; middle part of medium- to dark-gray thick-bedded dolomite

EROSIONAL UNCONFORMITY(?)

Mission Canyon limestone
Medium-gray to light-gray medium-grained thickly and indistinctly bedded limestone, with a few thin siliceous layers in lower 200 ft and sparse gray chert nodules and lentils in upper half. A breccia unit, br, about 200 ft below top of formation has been mapped locally

Lodgepole limestone
Upper part of medium-gray fine- to medium-grained limestone in distinct beds as much as 3 ft thick alternating with zones of much thinner beds containing rare mudstone partings; lower part of medium-gray limestone in beds 1 in. to 1 ft thick with partings and interbeds of yellow to red calcareous mudstone; grades into Mission Canyon limestone through a 150- to 200-ft zone

Three Forks shale
Predominantly greenish-gray and brown shale with subordinate amounts of interbedded sandstone and limestone. Dolomitic siltstone at top. Locally a 10- to 25-ft fossiliferous limestone unit, li, has been mapped

Jefferson dolomite
Dark-gray granular-weathering fetid well-bedded dolomite with subordinate amounts of dark-gray limestone and light-gray dolomite

EROSIONAL  *UNCONFORMITY*

Maywood and Red Lion formations undifferentiated,
Varicolored, generally in shades of red and yellowish-brown, argillaceous, dolomitic, and calcareous rocks; poorly exposed

Pilgrim dolomite
Comprises three units. Upper unit is light-gray thick-bedded dolomite commonly mottled medium-gray near base. Middle unit is light- to medium-gray crystalline limestone irregularly ribboned with yellowish-gray silty dolomite. Lower unit is mottled light- and dark-gray dolomite with sparse intraformational conglomerate; locally, basal 8 to 10 ft is bluish-gray limestone

Park shale
Olive-gray, gray, and light-brown shale with minor amounts of argillaceous limestone, siltstone, and sandstone

Meagher limestone
Comprises three units. Upper and lower units are medium-gray limestone irregularly ribboned or mottled with yellowish-orange, yellowish-brown, and yellowish-gray dolomite. Middle unit is thickly and indistinctly bedded medium-gray limestone, commonly with oolitic beds

Wolsey shale
Upper half is interbedded gray argillaceous limestone and greenish- and yellowish-gray calcareous mudstone and shale. Lower half is greenish-gray and drab shale with some interbeds of sandstone and limestone; many beds are micaceous, some are glauconitic

Flathead quartzite
White to pale shades of gray, pink, brown, and purple medium- to thick-bedded homogenous even-grained quartz sandstone; most beds are cemented to vitreous quartzite; thin, discontinuous sparse pebble zones in lower part; crossbedding common

UNCONFORMITY

Empire shale
Gray, greenish-gray, and brown, siliceous mudstone or argillite with interbeds of quartzite sandstone and shale. Intertongues with Spokane shale

Spokane shale
Grayish-red mudstone, shale, and sandstone, with a few thin beds of limestone near base

Greyson shale
Gray and brown mudstone or shale alternating with sandstone or quartzite. Base not exposed. Grades into Spokane shale

INTRUSIVE ROCKS

YOUNGER INTRUSIVE ROCKS

Granodiorite and quartz diorite
In Sagebrush Park stock

Composite or hybrid intrusives
Small plutons containing diverse and unusual rocks including olivine-rich and quartz-rich types

OLDER INTRUSIVE ROCKS

Basalt and related rocks
Dark-gray to greenish-black fine- to medium-grained rocks, mainly as sills

Andesite porphyry, diorite porphyry, and related rocks
Greenish-gray to dark-gray porphyritic rocks with phenocrysts of plagioclase and hornblende or augite; mainly as sills

Hornblende lamprophyre
Very fine grained gray rock with conspicuous hornblende phenocrysts

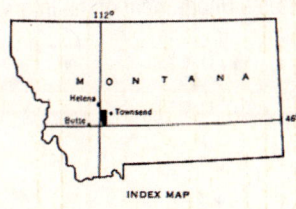

INDEX MAP

Earth Science Lab Manual ▪ DB9

Resource 8 Topographic Map of Campti, Louisiana

Source: United States Department of the Interior, Geological Survey

SCALE 1:62500
CONTOUR INTERVAL 20 FEET
DATUM IS MEAN SEA LEVEL

LOUISIANA
QUADRANGLE LOCATION

Earth Science Lab Manual • DB10

Resource 9: Topographic Map of Whitewater, Wisconsin

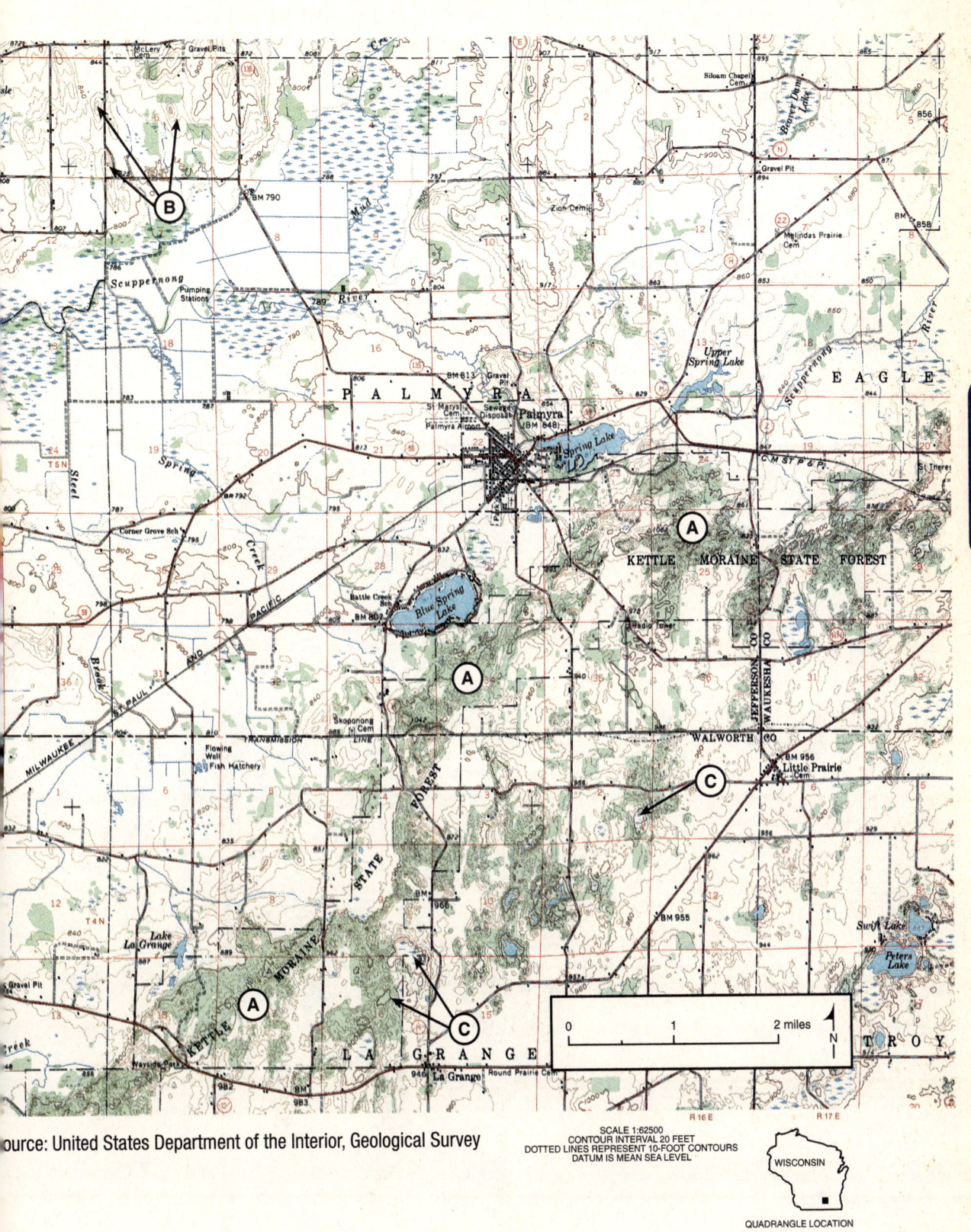

Source: United States Department of the Interior, Geological Survey

SCALE 1:62500
CONTOUR INTERVAL 20 FEET
DOTTED LINES REPRESENT 10-FOOT CONTOURS
DATUM IS MEAN SEA LEVEL

Resource 10 — The Geologic Time Scale

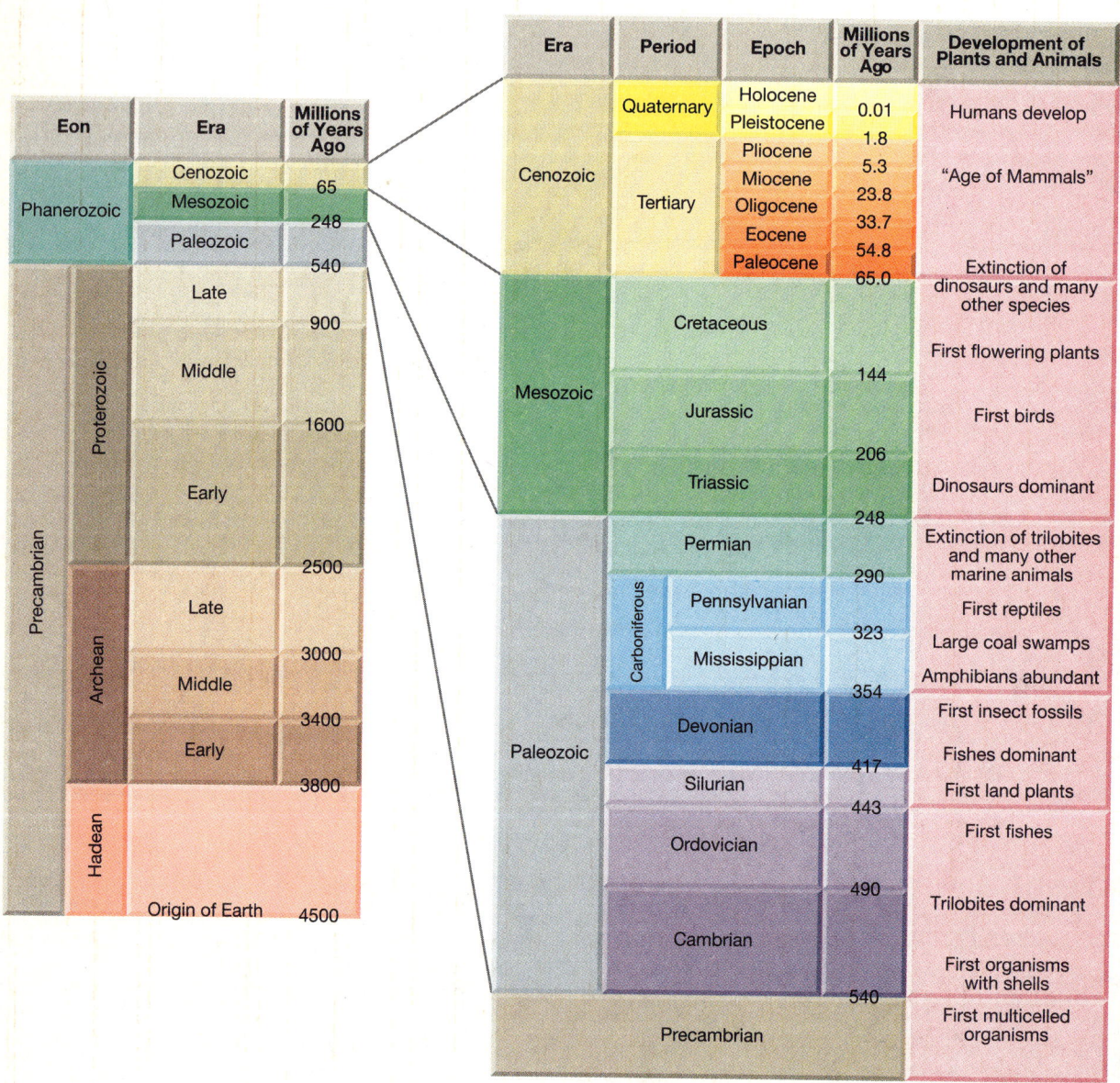

Resource 11: Key to Index Fossils

Era	Period	Index Fossil 1		Index Fossil 2	
CENOZOIC ERA (Age of Recent Life)	Quaternary Period	Pecten gibbus		Neptunea tabulata	
	Tertiary Period	Calyptraphorus velatus		Venericardia planicosta	
MESOZOIC ERA (Age of Medieval Life)	Cretaceous Period	Scaphites hippocrepis		Inoceramus labiatus	
	Jurassic Period	Perisphinctes tiziani		Nerinea trinodosa	
	Triassic Period	Trophites subbullatus		Monotis subcircularis	
PALEOZOIC ERA (Age of Ancient Life)	Permian Period	Leptodus americanus		Parafusulina bosei	
	Pennsylvanian Period	Dictyoclostus americanus		Lophophyllidium proliferum	
	Mississippian Period	Cactocrinus multibrachiatus		Prolecanites gurleyi	
	Devonian Period	Mucrospirifer mucronatus		Palmatolepus unicornis	
	Silurian Period	Cystiphyllum niagarense		Hexamoceras hertzeri	
	Ordovician Period	Bathyurus extans		Tetragraptus fructicosus	
	Cambrian Period	Paradoxides pinus		Billingsella corrugata	
PRECAMBRIAN	— — — —				

Earth Science Lab Manual

Resource 12: Atmospheric Temperature Curve

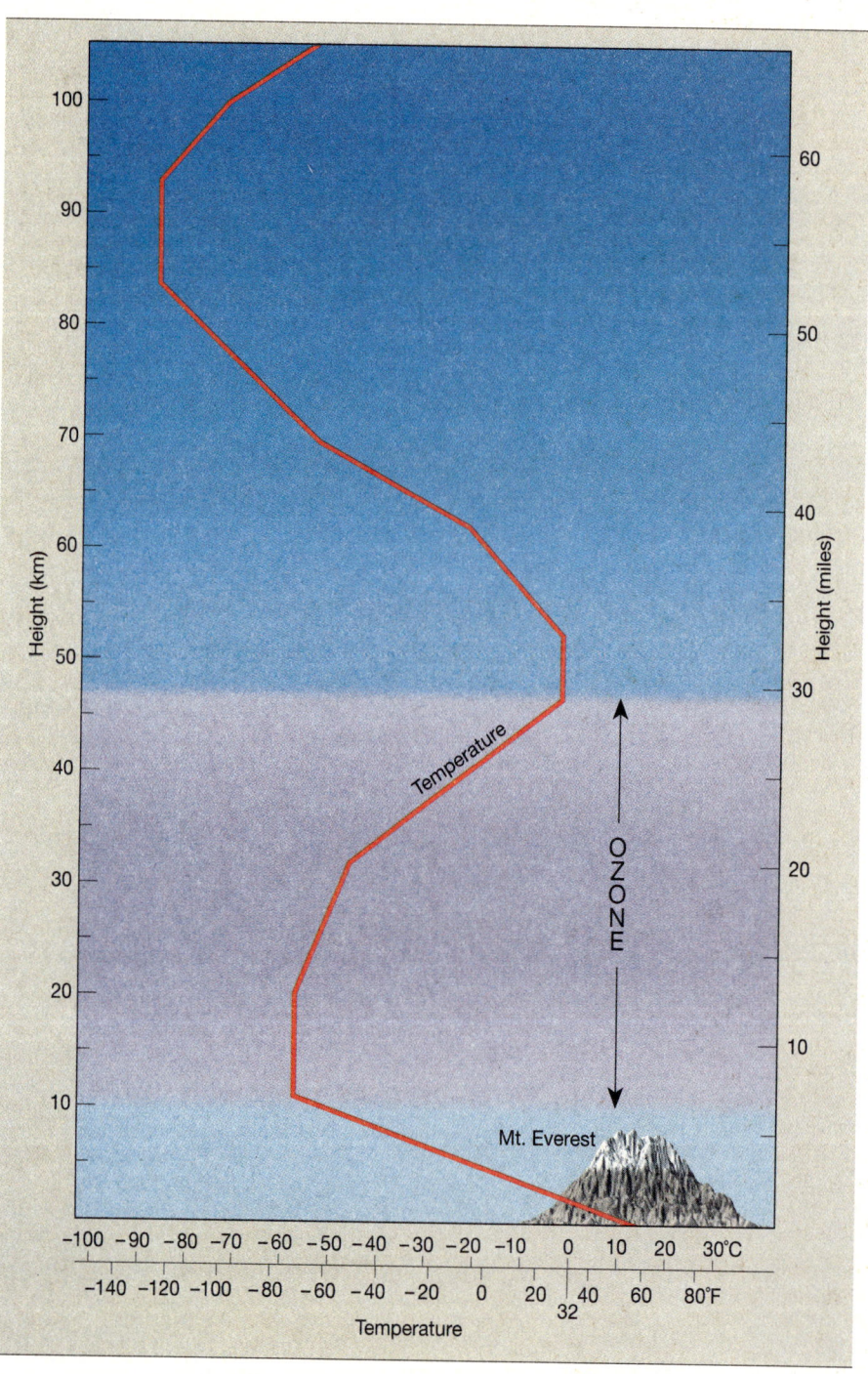

Earth Science Lab Manual • **DB14**

Resource 13 — Dew-Point Temperature Table

Dew-Point Temperature (°C)

(Dry-Bulb Temperature Minus Wet-Bulb Temperature = Depression of the Wet Bulb)

Dry Bulb (°C)	1	2	3	4	5	6	7	8	9	10	11	12	13	14	15	16	17	18	19	20	21	22
−20	−33																					
−18	−28																					
−16	−24																					
−14	−21	−36																				
−12	−18	−28																				
−10	−14	−22																				
−8	−12	−18	−29																			
−6	−10	−14	−22																			
−4	−7	−12	−17	−29																		
−2	−5	−8	−13	−20																		
0	−3	−6	−9	−15	−24																	
2	−1	−3	−6	−11	−17																	
4	1	−1	−4	−7	−11	−19																
6	4	1	−1	−4	−7	−13	−21															
8	6	3	1	−2	−5	−9	−14															
10	8	6	4	1	−2	−5	−9	−14	−18													
12	10	8	6	4	1	−2	−5	−9	−16													
14	12	11	9	6	4	1	−2	−5	−10	−17												
16	14	13	11	9	7	4	1	−1	−6	−10	−17											
18	16	15	13	11	9	7	4	2	−2	−5	−10	−19										
20	19	17	15	14	12	10	7	4	2	−2	−5	−10	−19									
22	21	19	17	16	14	12	10	8	5	3	−1	−5	−10	−19								
24	23	21	20	18	16	14	12	10	8	6	2	−1	−5	−10	−18							
26	25	23	22	20	18	17	15	13	11	9	6	3	0	−4	−9	−18						
28	27	25	24	22	21	19	17	16	14	11	9	7	4	1	−3	−9	−16					
30	29	27	26	24	23	21	19	18	16	14	12	10	8	5	1	−2	−8	−15				
32	31	29	28	27	25	24	22	21	19	17	15	13	11	8	5	2	−2	−7	−14			
34	33	31	30	29	27	26	24	23	21	20	18	16	14	12	9	6	3	−1	−5	−12	−29	
36	35	33	32	31	29	28	27	25	24	22	20	19	17	15	13	10	7	4	0	−4	−10	
38	37	35	34	33	32	30	29	28	26	25	23	21	19	17	15	13	11	8	5	1	−3	9
40	39	37	36	35	34	32	31	30	28	27	25	24	22	20	18	16	14	12	9	6	2	−2

Dry-Bulb (Air) Temperature

Dew-Point Values

Resource 14 — Temperature Contour Plots

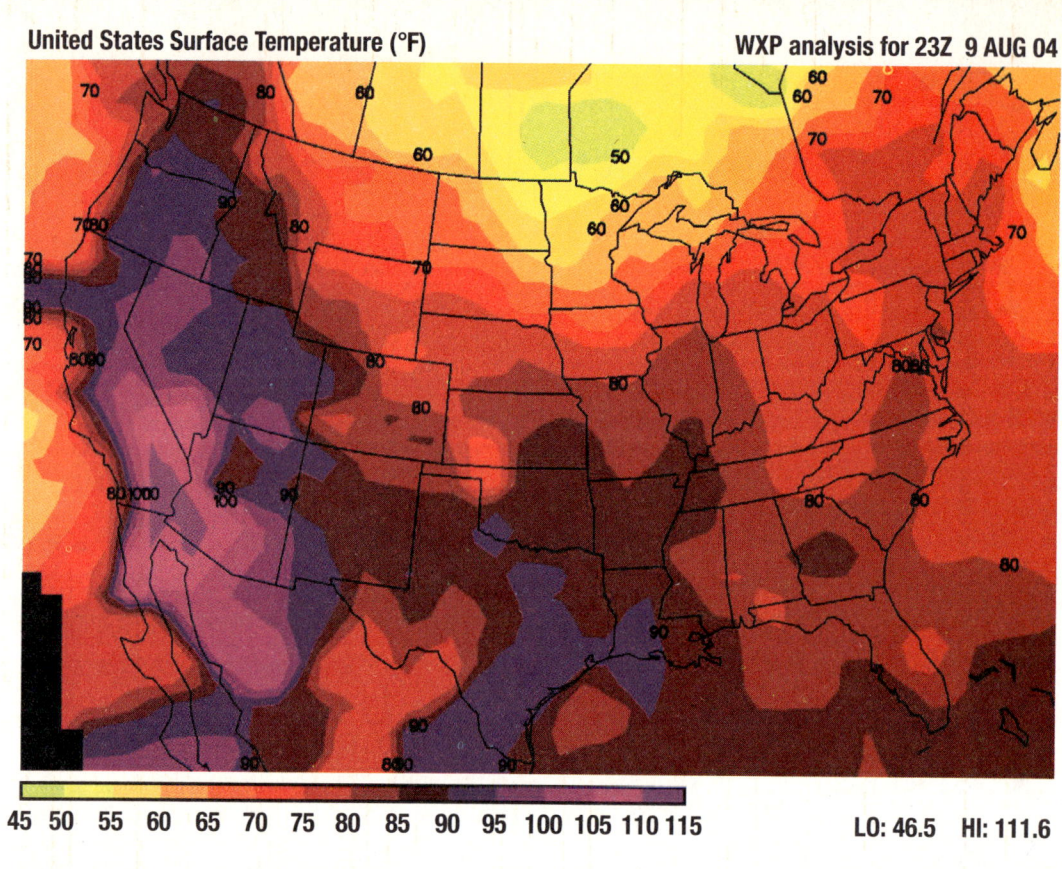

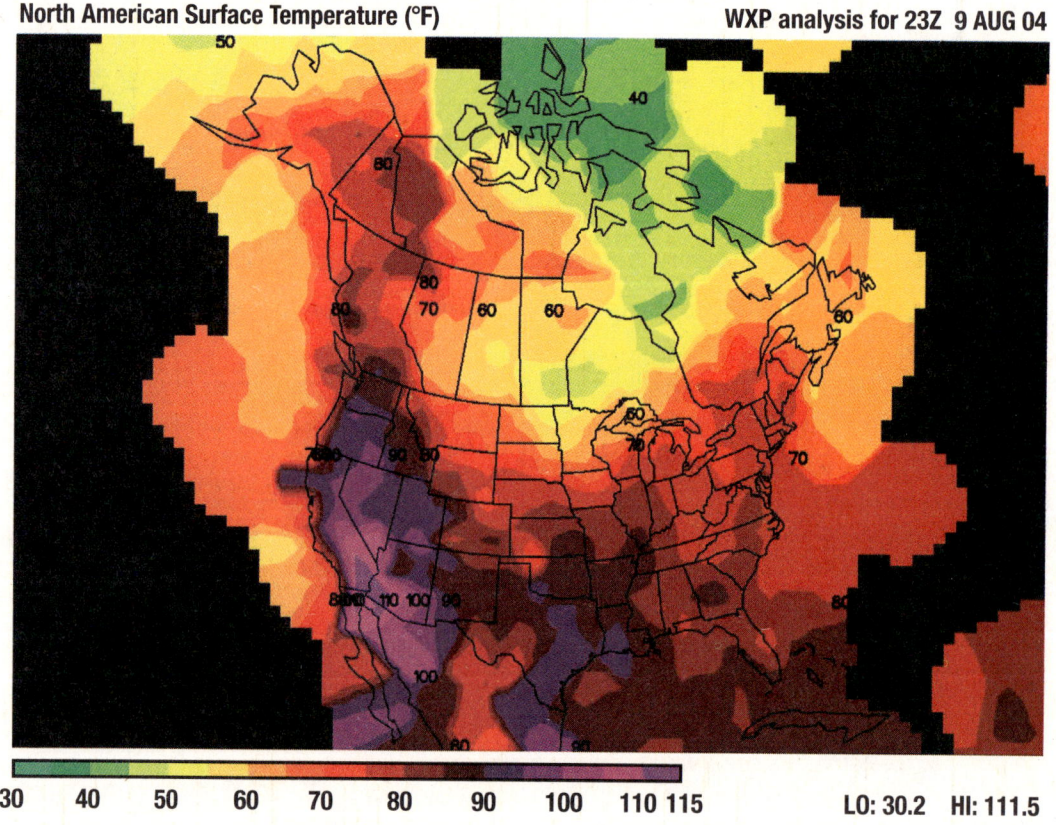

Earth Science Lab Manual ▪ **DB16**

Resource 15: Temperature Change and Heat Index Plots

24-Hour Temperature Change (°F) — 23Z 9 AUG 04

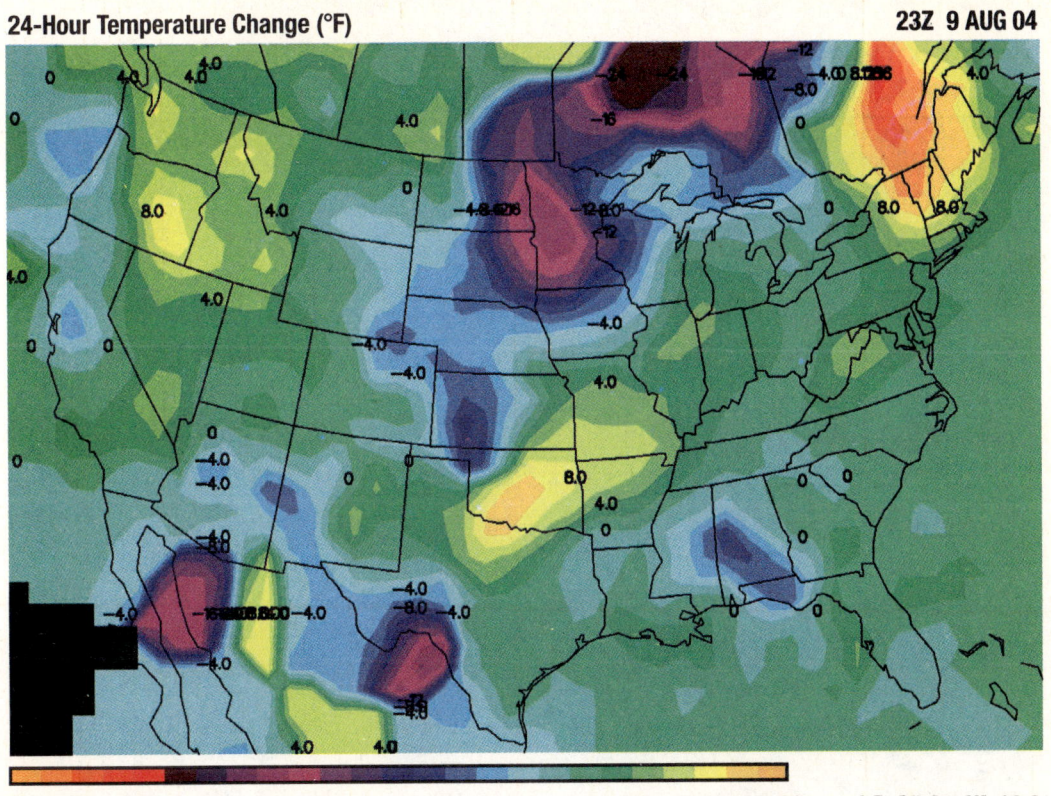

45 −32 −28 −24 −20 −16 −12 −8.0 −4.0 0 4.0 8.0 12 14   LO: 25.2   HI: 16.9

Surface Heat Index (°F) — WXP analysis for 23Z 9 AUG 04

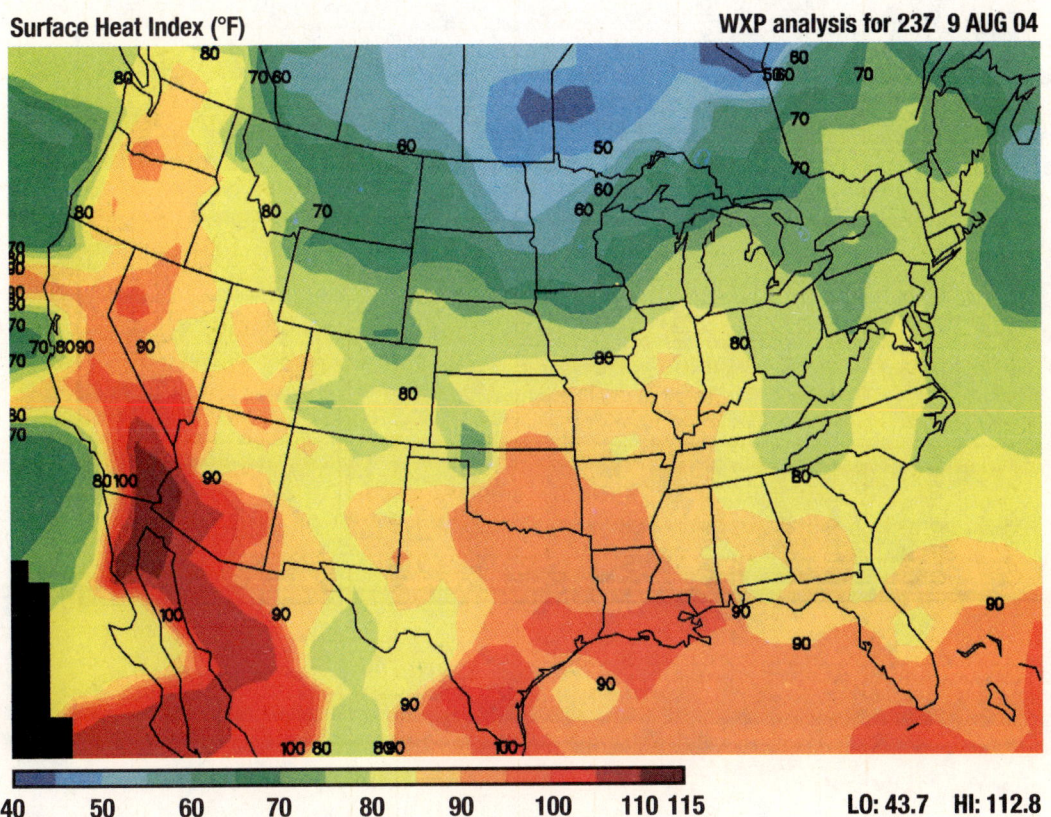

40 50 60 70 80 90 100 110 115   LO: 43.7   HI: 112.8

Earth Science Lab Manual ■ DB17

Resource 16: Some Common Minerals and Their Properties

Name	Chemical Formula and Mineral Group	Common Color(s)	Density (g/cm³)	Hardness	Comments
Quartz	SiO_2 silicates	colorless, milky white, pink, brown	2.65	7	glassy luster; conchoidal fractures
Orthoclase feldspar	$KAlSi_3O_8$ silicates	white to pink	2.57	6	cleaves in two directions at 90°
Plagioclase feldspar	$(Na,Ca)AlSi_3O_8$ silicates	white to gray	2.69*	6	cleaves in two directions at 90°; striations common
Galena	PbS sulfides	metallic silver	7.5*	2.5	cleaves in two directions at 90°; lead gray streak
Pyrite	FeS_2 sulfides	brassy yellow	5.02	6–6.5	fractures; forms cubic crystals; greenish-black streak
Sulfur	S native elements	yellow	2.07*	1.5–2.5	fractures; yellow streak smells like rotten eggs
Fluorite	CaF_2 halides	colorless, purple	3.18	4	perfect cleavage in four directions; glassy luster
Olivine	$(Mg,Fe)_2SiO_4$ silicates	green, yellowish-green	3.82*	6.5–7	fractures; glassy luster; often has granular texture
Calcite	$CaCO_3$ carbonates	colorless, gray	2.71	3	bubbles with HCl; cleaves in three directions
Talc	$Mg_3Si_4O_{10}(OH)_2$ silicates	pale green, gray, white	2.75*	1	pearly luster; feels greasy; cleaves in one direction
Gypsum	$CaSO_4 \cdot 2H_2O$ sulfates	colorless, white, gray	2.32	2	glassy or pearly luster; cleaves in three directions
Muscovite mica	$KAl_3Si_3O_{10}(OH)_2$ silicates	colorless in thin sheets to brown	2.82*	2–2.5	silky to pearly luster; cleaves in one direction to form flexible sheets

* Average density of the mineral

Name	Chemical Formula and Mineral Group	Common Color(s)	Density (g/cm³)	Hardness	Comments
Biotite mica	$K(Mg,Fe)_3(AlSi_3O_{10})(OH)_2$ silicates	dark green to brown to black	3.0*	2.5–3	perfect cleavage in one direction to form flexible sheets
Halite	$NaCl$ halides	colorless, white	2.16	2.5	has a salty taste; dissolves in water; cleaves in three directions
Augite	$(Ca, Na)(Mg, Fe, Al)(Si, Al)_2O_6$ silicates	dark green to black	3.3*	5–6	glassy luster; cleaves in two directions; crystals have 8-sided cross section
Hornblende	$(Ca, Na)_{2-3}(MgFeAl)_5Si_6(SiAl)_2O_{22}(OH)_2$ silicates	dark green to black	3.2*	5–6	glassy luster; cleaves in two directions; crystals have 6-sided cross section
Hematite	Fe_2O_3 oxides	reddish brown to black	5.26	5.5–6.5	metallic luster in crystals; dull luster in earthy variety; dark red streak
Dolomite	$CaMg(CO_3)_2$ carbonates	pink, colorless, white, gray	2.85	3.5–4	does not react to HCl as quickly as calcite; cleaves in three directions
Magnetite	Fe_3O_4 oxides	black	5.18	6	metallic luster; black streak; strongly magnetic
Copper	Cu native elements	copper-red on fresh surface	8.9	2.5–3	metallic luster; fractures; can be easily shaped
Graphite	C native elements	black to gray	2.3	1–2	black to gray streak; marks paper; feels slippery

Resource 17: Classification of Rocks

Classification of Major Igneous Rocks

			Granitic	Andesitic	Basaltic	Ultramafic
	Chemical Composition					
	Dominant Minerals		Quartz, Potassium feldspar, Sodium-rich plagioclase feldspar	Amphibole, Sodium- and calcium-rich plagioclase feldspar	Pyroxene, Calcium-rich plagioclase feldspar	Olivine, Pyroxene
TEXTURE	Coarse-grained		Granite	Diorite	Gabbro	Peridotite
	Fine-grained		Rhyolite	Andesite	Basalt	Komatiite (rare)
	Porphyritic		"Porphyritic" precedes any of the above names whenever there are appreciable phenocrysts.			Uncommon
	Glassy		Obsidian (compact glass) / Pumice (frothy glass)			
	Rock Color (based on % of dark minerals)		0% to 25%	25% to 45%	45% to 85%	85% to 100%

Classification of Major Metamorphic Rocks

Rock Name		Texture	Grain Size	Comments	Parent Rock
Slate	Increasing Metamorphism ↓	Foliated	Very fine	Smooth dull surfaces	Shale, mudstone, or siltstone
Phyllite		Foliated	Fine	Breaks along wavey surfaces, glossy sheen	Slate
Schist		Foliated	Medium to Coarse	Micaceous minerals dominate	Phyllite
Gneiss		Foliated	Medium to Coarse	Banding of minerals	Schist, granite, or volcanic rocks
Marble		Nonfoliated	Medium to coarse	Interlocking calcite or dolomite grains	Limestone, dolostone
Quartzite		Nonfoliated	Medium to coarse	Fused quartz grains, massive, very hard	Quartz sandstone
Anthracite		Nonfoliated	Fine	Shiny black organic rock that fractures	Bituminous coal

Classification of Major Sedimentary Rocks

Clastic Sedimentary Rocks

Texture (grain size)	Sediment Name	Rock Name
Coarse (over 2 mm)	Gravel (rounded fragments)	Conglomerate
Coarse (over 2 mm)	Gravel (angular fragments)	Breccia
Medium (1/16 to 2 mm)	Sand	Sandstone
Fine (1/16 to 1/256 mm)	Mud	Siltstone
Very fine (less than 1/256 mm)	Mud	Shale

Chemical Sedimentary Rocks

Composition	Texture (grain size)	Rock Name	
Calcite, $CaCO_3$	Fine to coarse crystalline	Crystalline Limestone	
Calcite, $CaCO_3$	Fine to coarse crystalline	Travertine	
Calcite, $CaCO_3$	Visible shells and shell fragments loosely cemented	Coquina	Biochemical Limestone
Calcite, $CaCO_3$	Various size shells and shell fragments cemented with calcite cement	Fossiliferous Limestone	Biochemical Limestone
Calcite, $CaCO_3$	Microscopic shells and clay	Chalk	Biochemical Limestone
Quartz, SiO_2	Very fine crystalline	Chert (light colored) Flint (dark colored)	
Gypsum $CaSO_4 \cdot 2H_2O$	Fine to coarse crystalline	Rock Gypsum	
Halite, NaCl	Fine to coarse crystalline	Rock Salt	
Altered plant fragments	Fine-grained organic matter	Bituminous Coal	

DataBank

Earth Science Lab Manual

Resource 18: Topography of the Ocean Floor

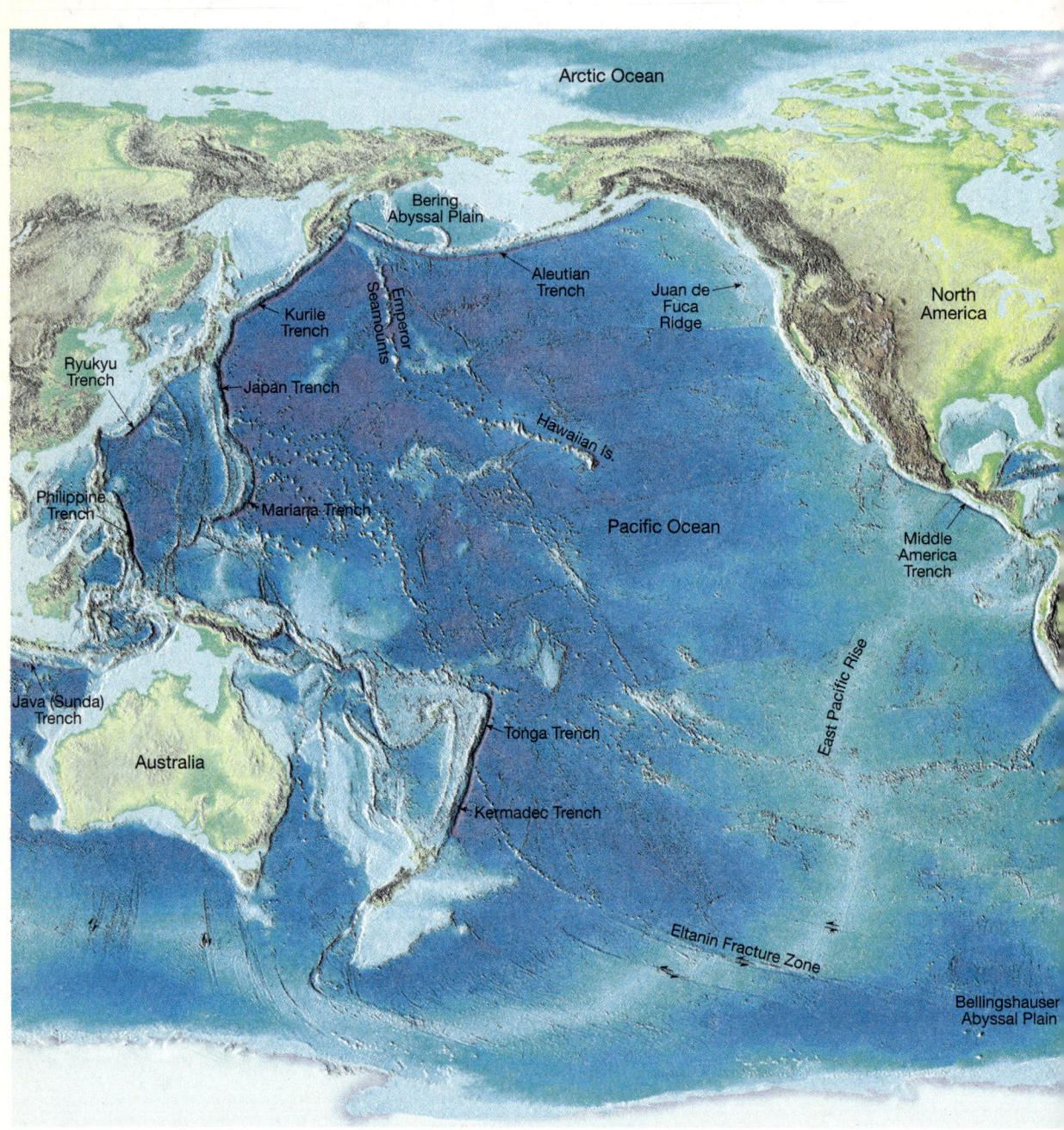

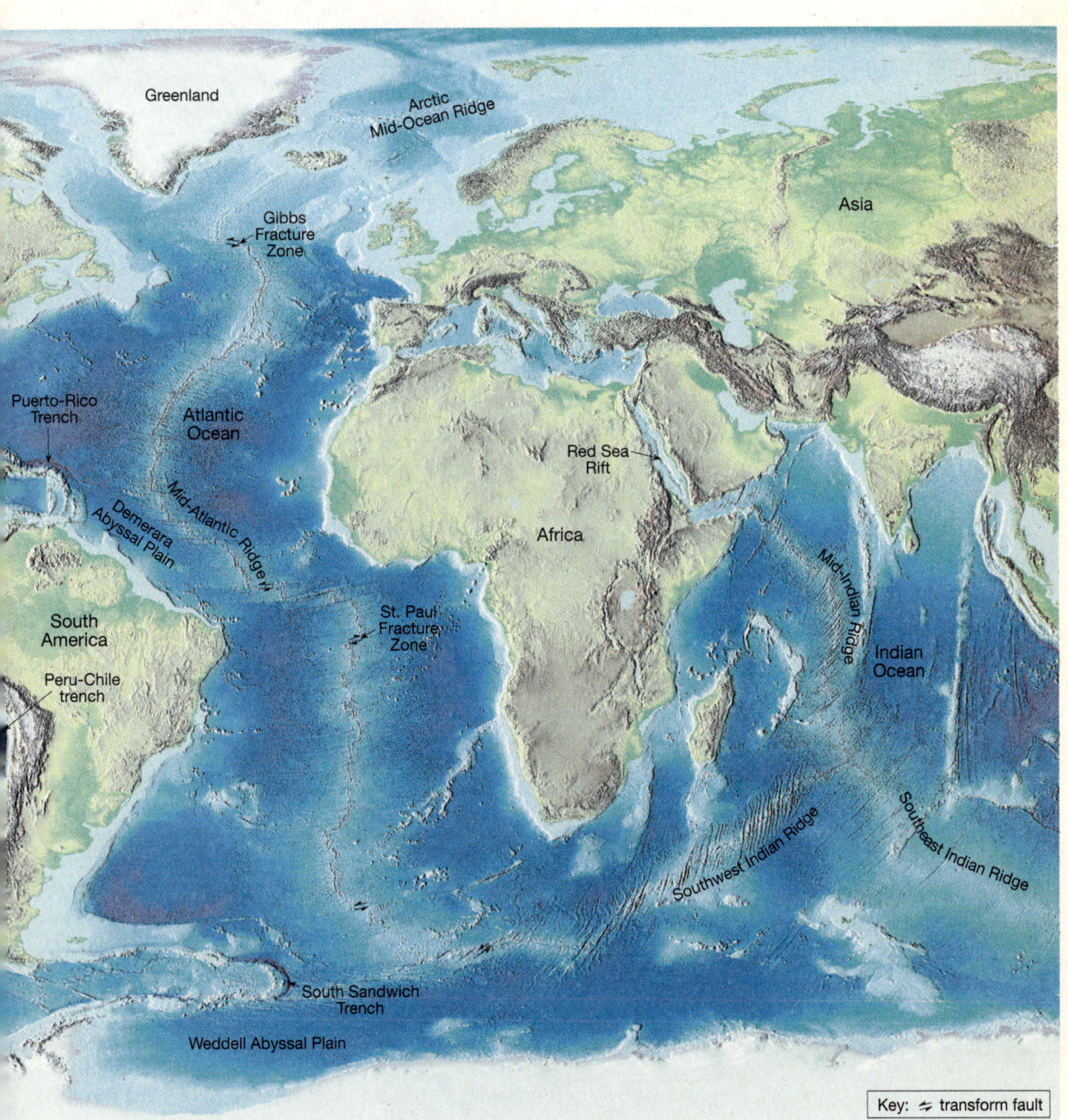

Resource 19: Star Charts

Spring Sky

To use this chart, hold it up in front of you and turn it so the direction you are facing is at the bottom of the chart. The chart works best at 35° N latitude, but it can be used at other latitudes. It works best at the following dates and times: March 1 at 10 P.M. and April 1 at 8 P.M.

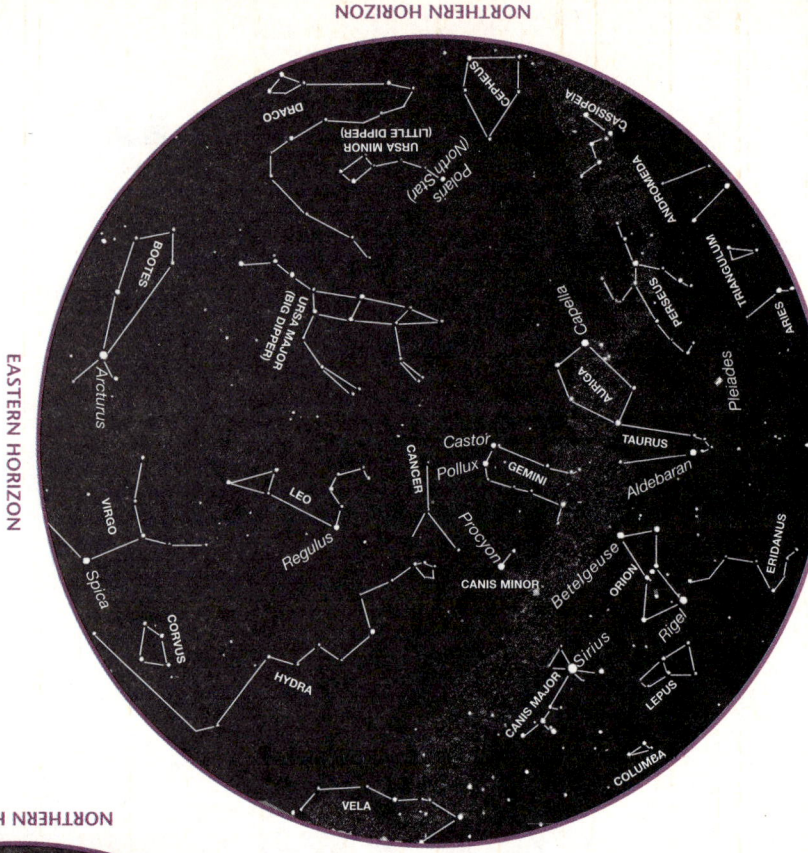

Summer Sky

To use this chart, hold it up in front of you and turn it so the direction you are facing is at the bottom of the chart. The chart works best at 35° N latitude, but it can be used at other latitudes. It works best at the following dates and times: May 15 at 11 P.M. and June 15 at 9 P.M.

Earth Science Lab Manual

Autumn Sky

To use this chart, hold it up in front of you and turn it so the direction you are facing is at the bottom of the chart. The chart works best at 35° N latitude, but it can be used at other latitudes. It works best at the following dates and times: September 1 at 10 P.M., October 1 at 8 P.M., and November 1 at 6 P.M.

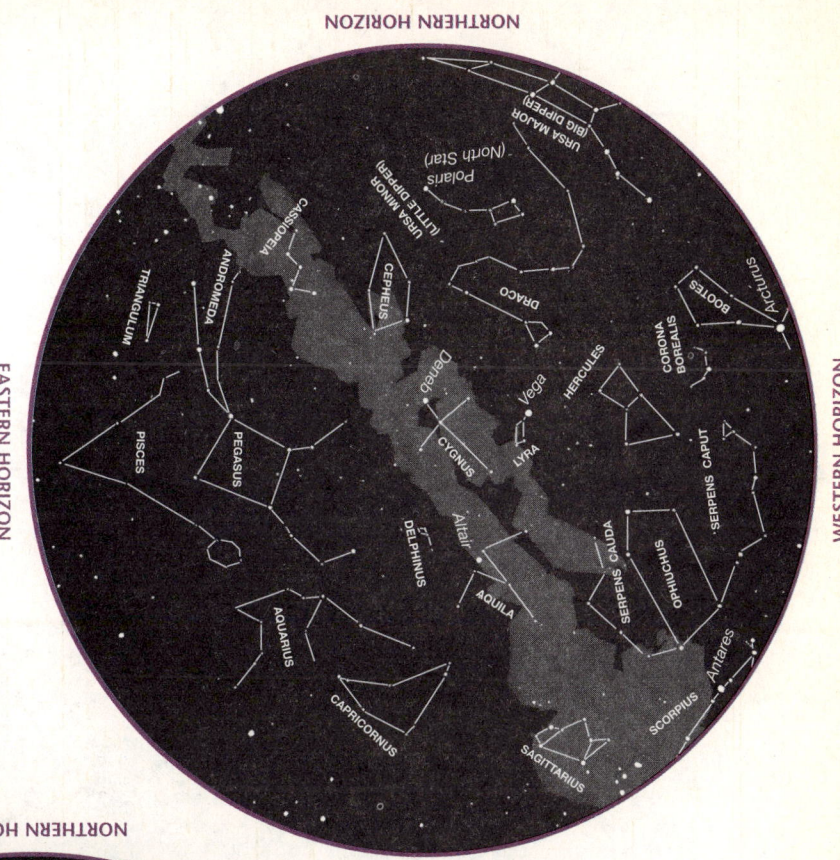

Winter Sky

To use this chart, hold it up in front of you and turn it so the direction you are facing is at the bottom of the chart. The chart works best at 35° N latitude, but it can be used at other latitudes. It works best at the following dates and times: December 1 at 10 P.M., January 1 at 8 P.M., and February 1 at 6 P.M.

Resource 20: Landforms of the Conterminous United States

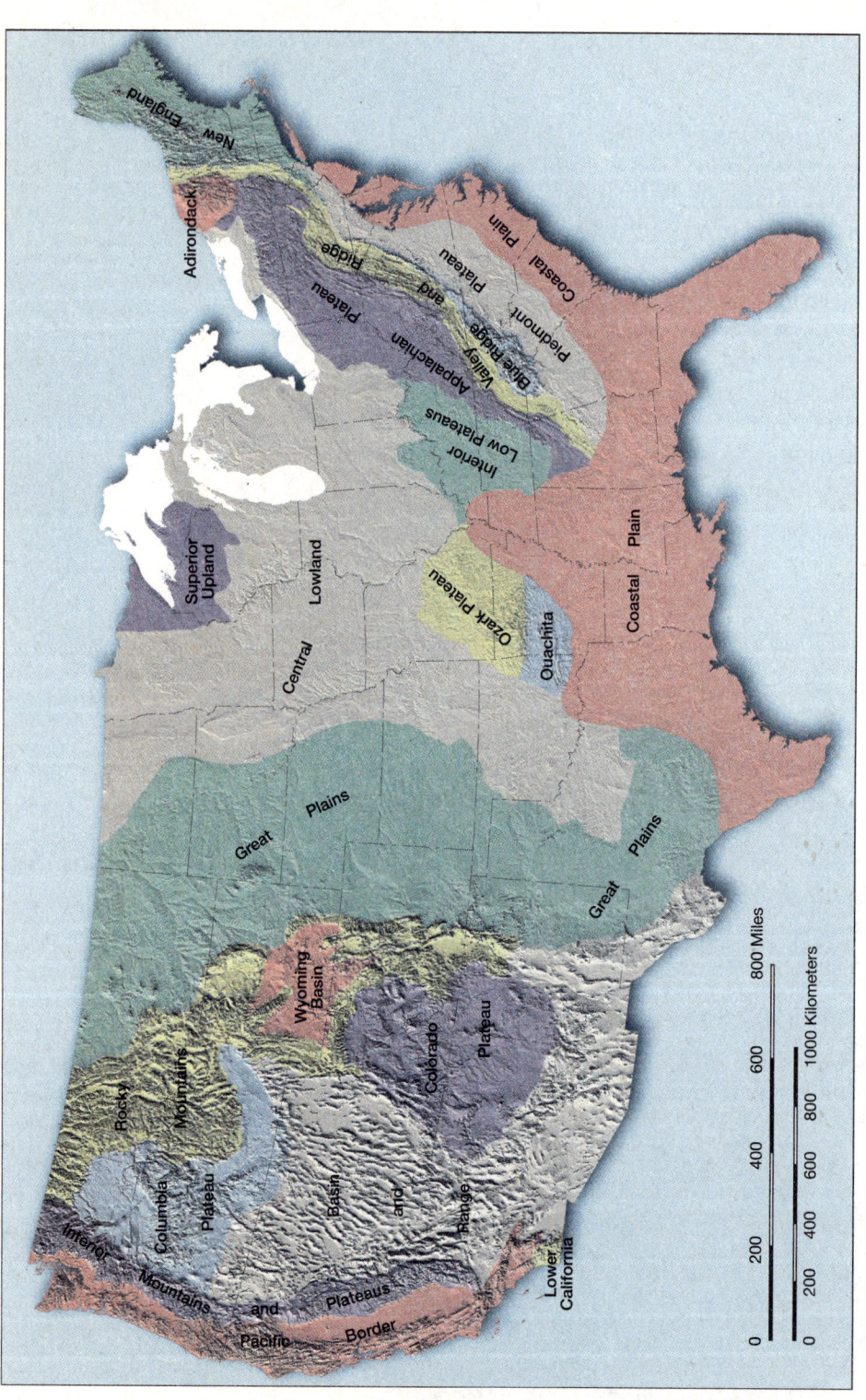

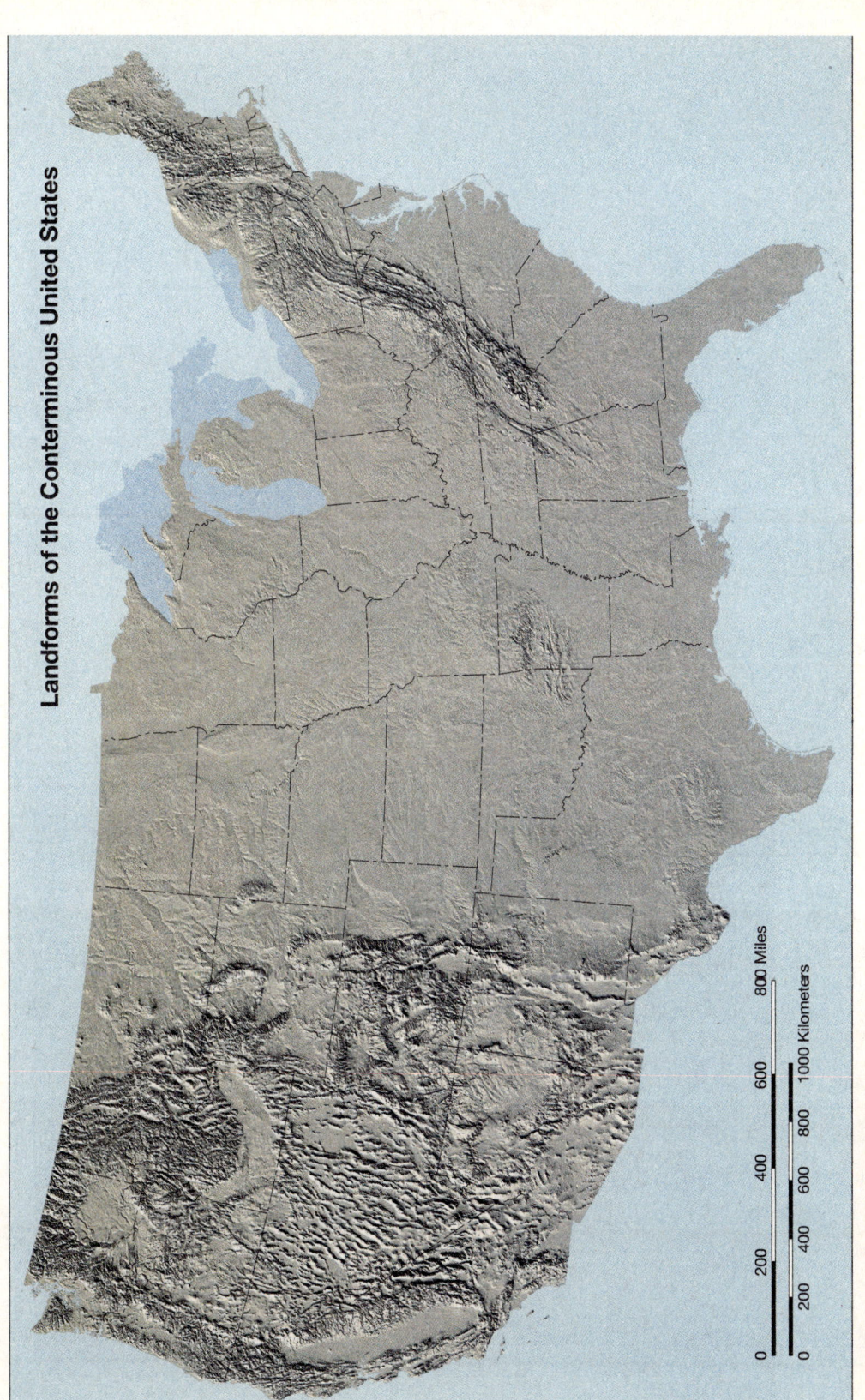

Resource 21: Circulation on a Rotating Earth

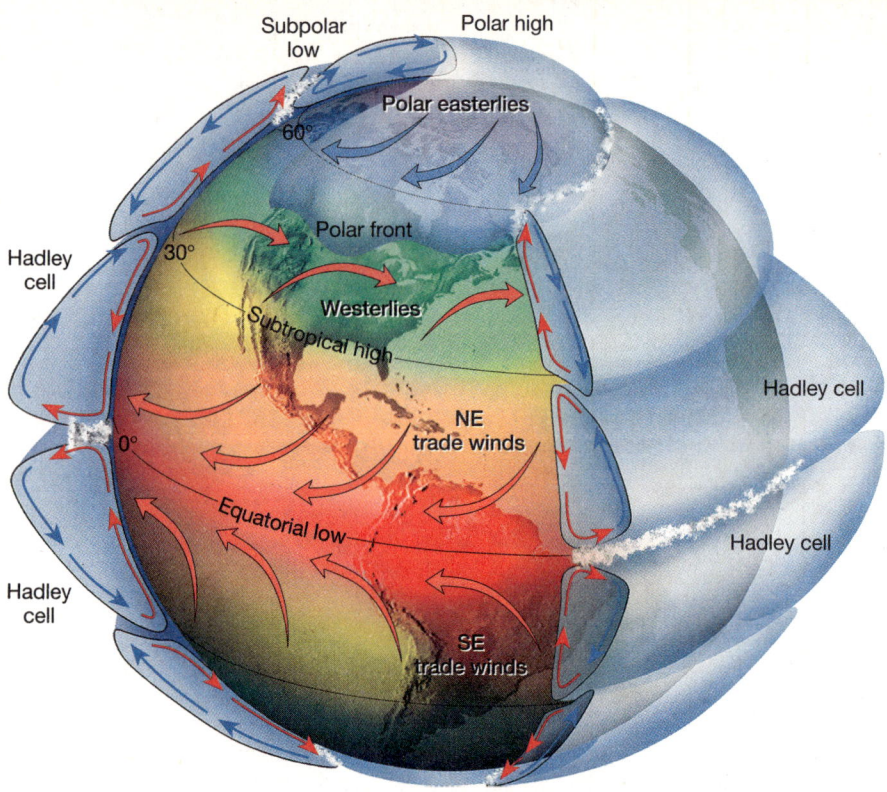

Hertzsprung-Russel Diagram

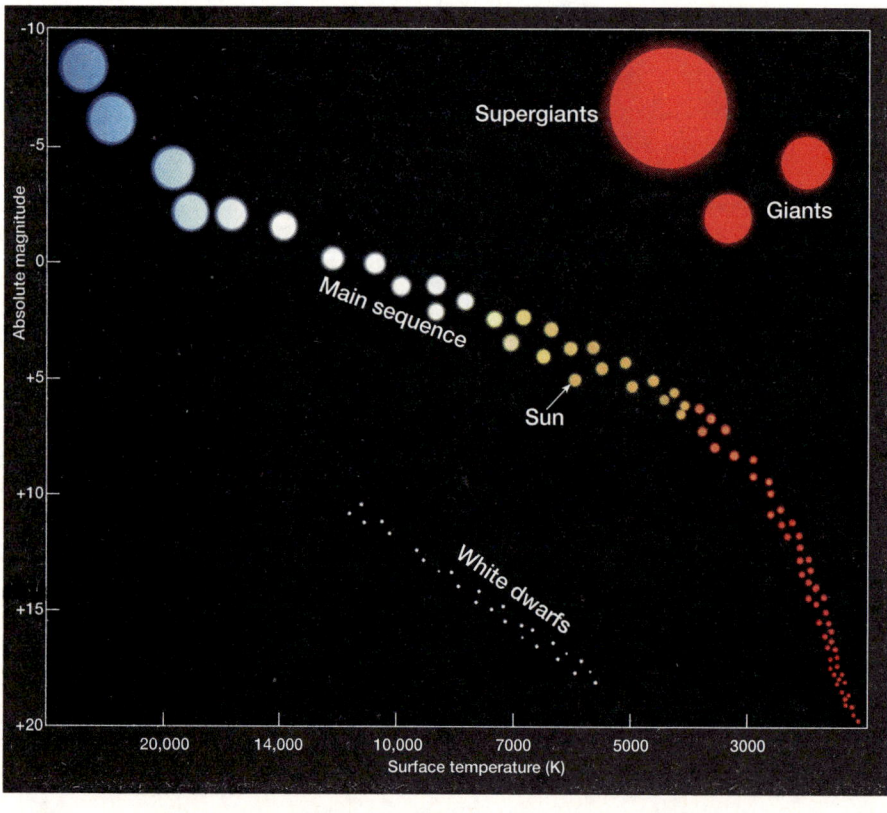

Resource 22: Middle-Latitude Cyclone Model

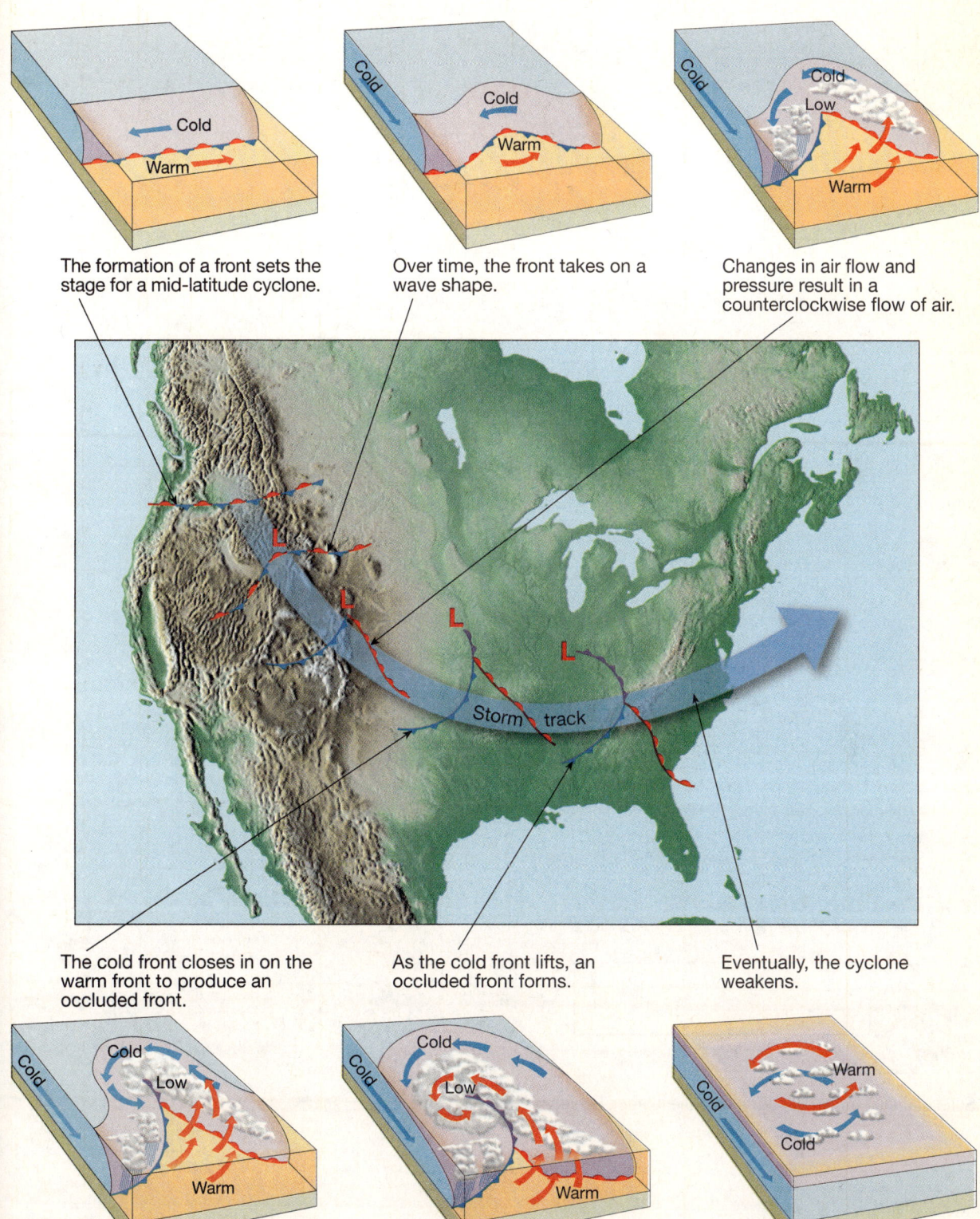

The formation of a front sets the stage for a mid-latitude cyclone.

Over time, the front takes on a wave shape.

Changes in air flow and pressure result in a counterclockwise flow of air.

The cold front closes in on the warm front to produce an occluded front.

As the cold front lifts, an occluded front forms.

Eventually, the cyclone weakens.

Earth Science Lab Manual • DB29

Resource 23: Topographic Map of a Glacial Landscape

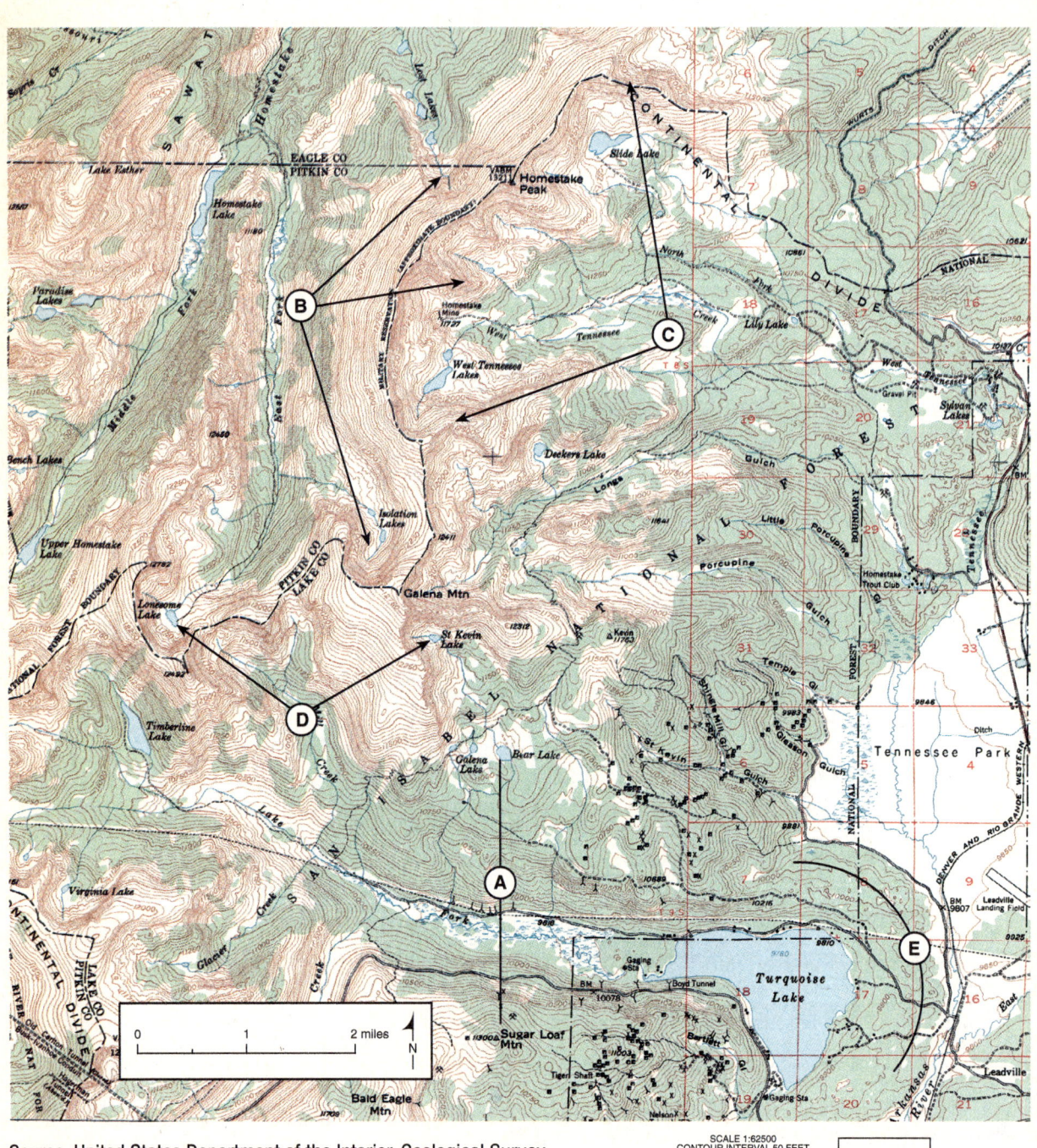

Source: United States Department of the Interior, Geological Survey

SCALE 1:62500
CONTOUR INTERVAL 50 FEET
DATUM IS MEAN SEA LEVEL

QUADRANGLE LOCATION: COLORADO